How still,
How strangely still
The water is today.

It is not good
For water
To be so still that way.

Langston HUGHES SEA CALM 1934

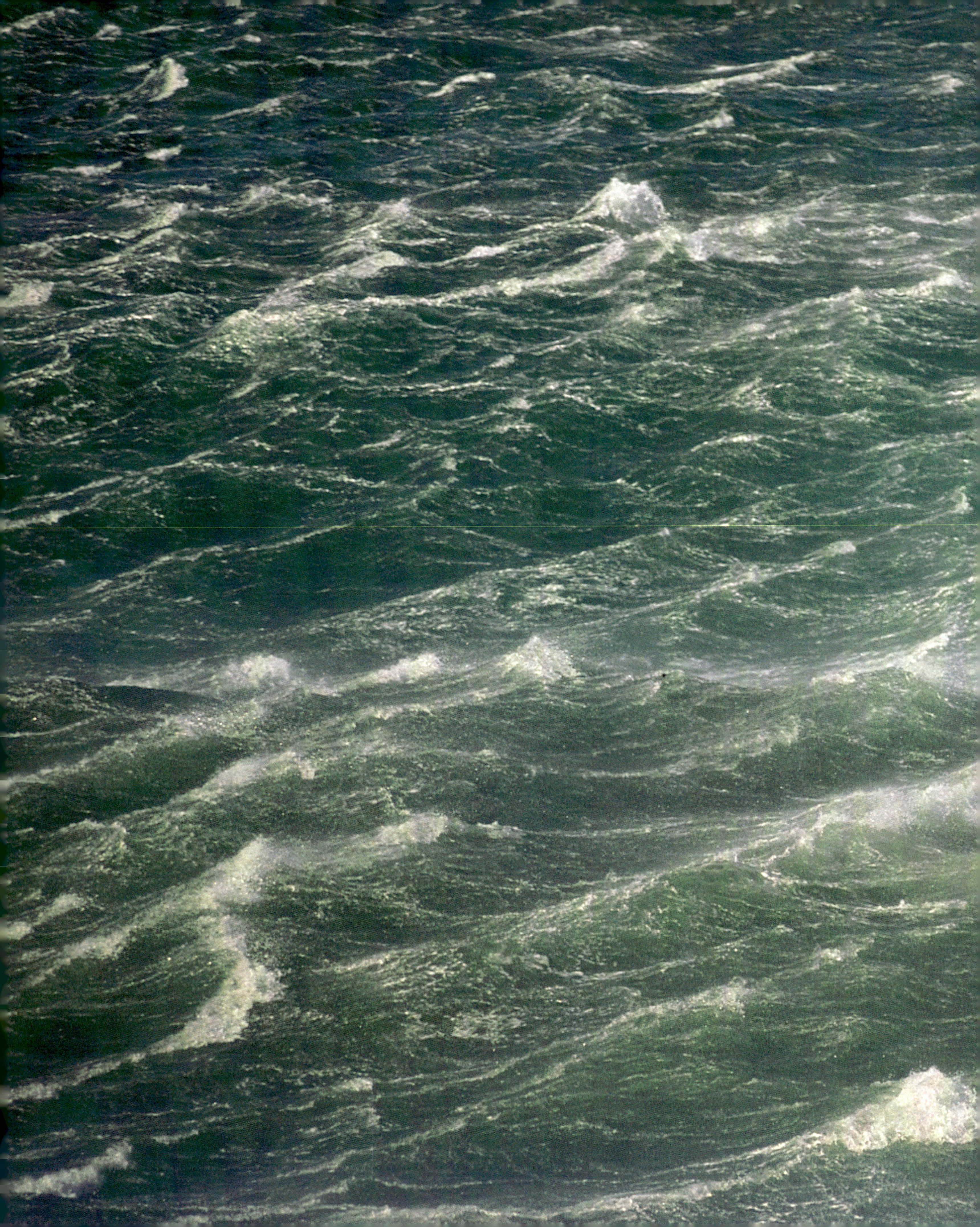

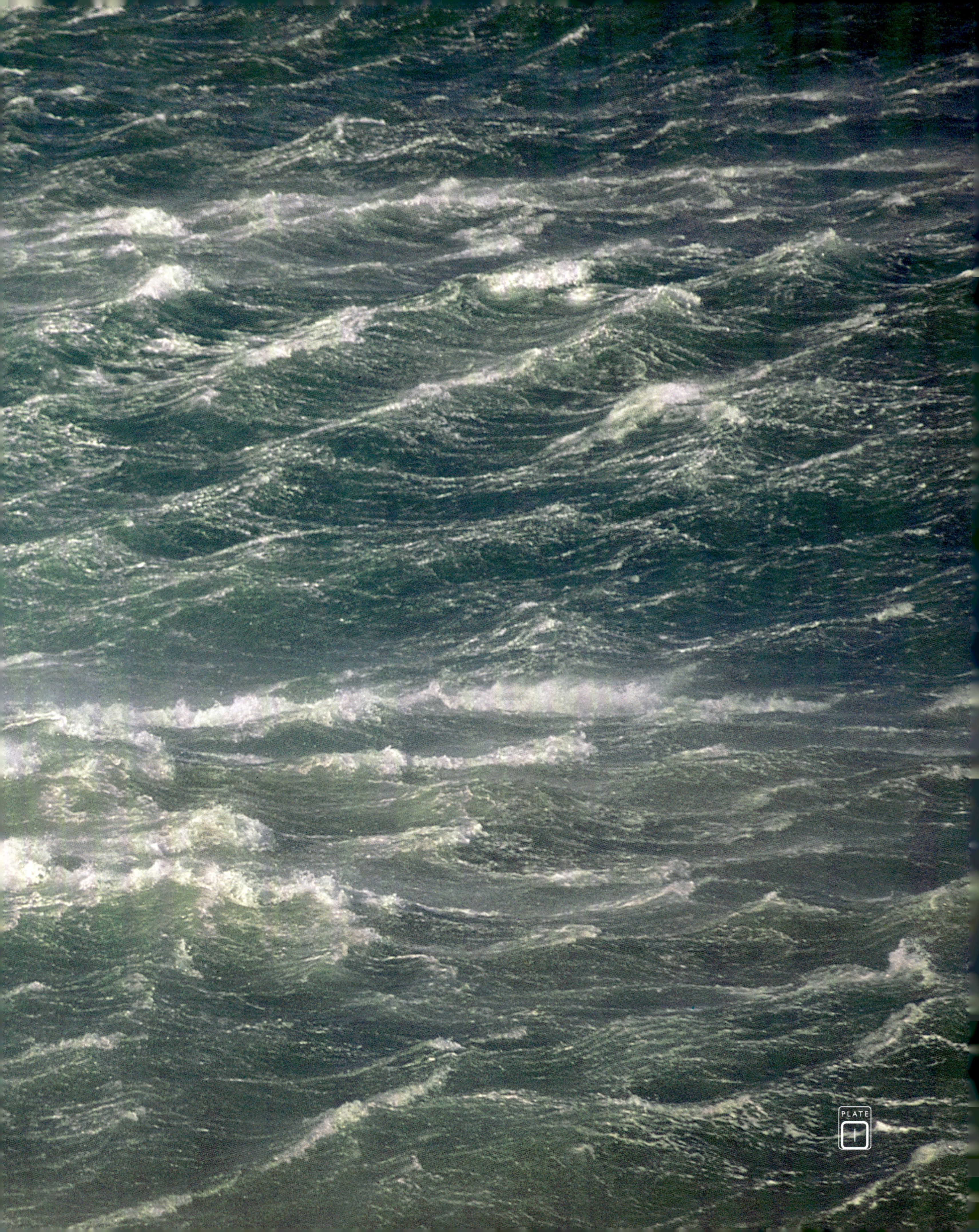
PLATE 1

introduction

Waves, in all their forms, sizes and shapes, have long exerted an extraordinary grip on the human imagination. Symbols of beauty to the Romantic poets, harbingers of fear and death to Atlantic fishermen, playthings and navigational tools of the Polynesians, they pour endlessly onto the coasts of the planet, in an easy expression of natural power both new and many hundreds of millions of years old.

Part of the breaking wave's hold on us is a simple matter of aesthetics: the graceful play of light and shadow across a liquid surface at once permanent and in constant change. Part is surely a matter of half-comprehending awe at the energies set loose by a large storm surf, for instance, or a massive long groundswell hitting full force along a reef or coastline. But part, too, is the lure of the coast itself — a place where the two worlds of land and sea come together to battle, or to broker some uneasy truce in a never-ending war. Much of the last fifty years of human affairs have been about coasts of one kind or another: the World War II coasts of Normandy, Iwo Jima, Guadalcanal and Pearl Harbour; the artificial coastline of the Berlin Wall; the American West Coast and its monumental leaps through the information age; and today, the coastal barriers that hold back tides of refugees, the coastal paradises that lure floods of tourists. Empowered by the wealth of booming Western economies, people are more and more drawn to coasts where real estate values spiral, driven ever higher by the promise of an ocean view. We stroll shores along which are spread all the evidence of the land-sea duel: shattered shells, bones of fish and empty stingray eggs, driftwood and dried kelp, and all the ground-up garbage that finds its way these days from our homes and factories onto the beaches of the world. Each tide brings to the shore more remnants of lives lived elsewhere; each big swell washes it clean. Perhaps something in these simple cycles, in the motion of waves and tides along the coast, brings a kind of peace to the human heart.

Meanwhile, far over the horizon, enormous natural forces are at work. Ocean waves are generally the result of the friction of moving air on the surface of the sea, and their creation follows a distinct and clearly marked pattern. With the first breath of wind come capillary waves less than a millimetre in height, existing on the fringe of surface tension, providing — however small — a tilt against which the wind can work. Then follow wind waves, thin, high and unstable, crumbling and falling over each other, yet always tending toward some kind of order and structure and flow. If the wind blows long enough over a broad enough stretch of water, the result will be a fully developed sea state, in which wind waves begin to group together and form long trains, bands of energy capable of moving far beyond the windy region of their birth and travelling thousands of kilometres of open ocean in an ever-smoothing sine wave of lifting water we know as swell. The energy exchange between air and water in this wave-making process is all in favour of the water, and when deposited it's capable of terrifying acts of power: ten-thousand-tonne rock masses have been known to shift and break free of their moorings under a storm surf's impact, and any coast-dweller will have witnessed the speed at which a strong swell can rip dirt, rock, sand, and anything else in its path away from the shore. Scientists have made astonishing estimates of wave energy on beaches, and there's no doubt that at any one time, more power is roaming the oceans of the world in the form of swell than we are able to harness in all our human endeavours. Yet, aside from surfers and the South Pacific navigators who once steered their double-hulled canoes by sensing the deep-ocean swell trends, humans have found no way to get hold of such energies and turn them to our use; they travel the globe, power incarnate, fettered only by their meetings with the shore, constantly reborn by the wind. Perhaps in the end, that is what we find most alluring about the sight of a wave rising up to break on a reef or beach — in a world where so many things seem somehow cramped, to shrink under our sway, still the waves are free.

The flood lasted forty days on the earth. The waters swelled, lifting the ark until it was raised above the earth. The waters rose and swelled greatly on the earth, and the ark sailed on the water..

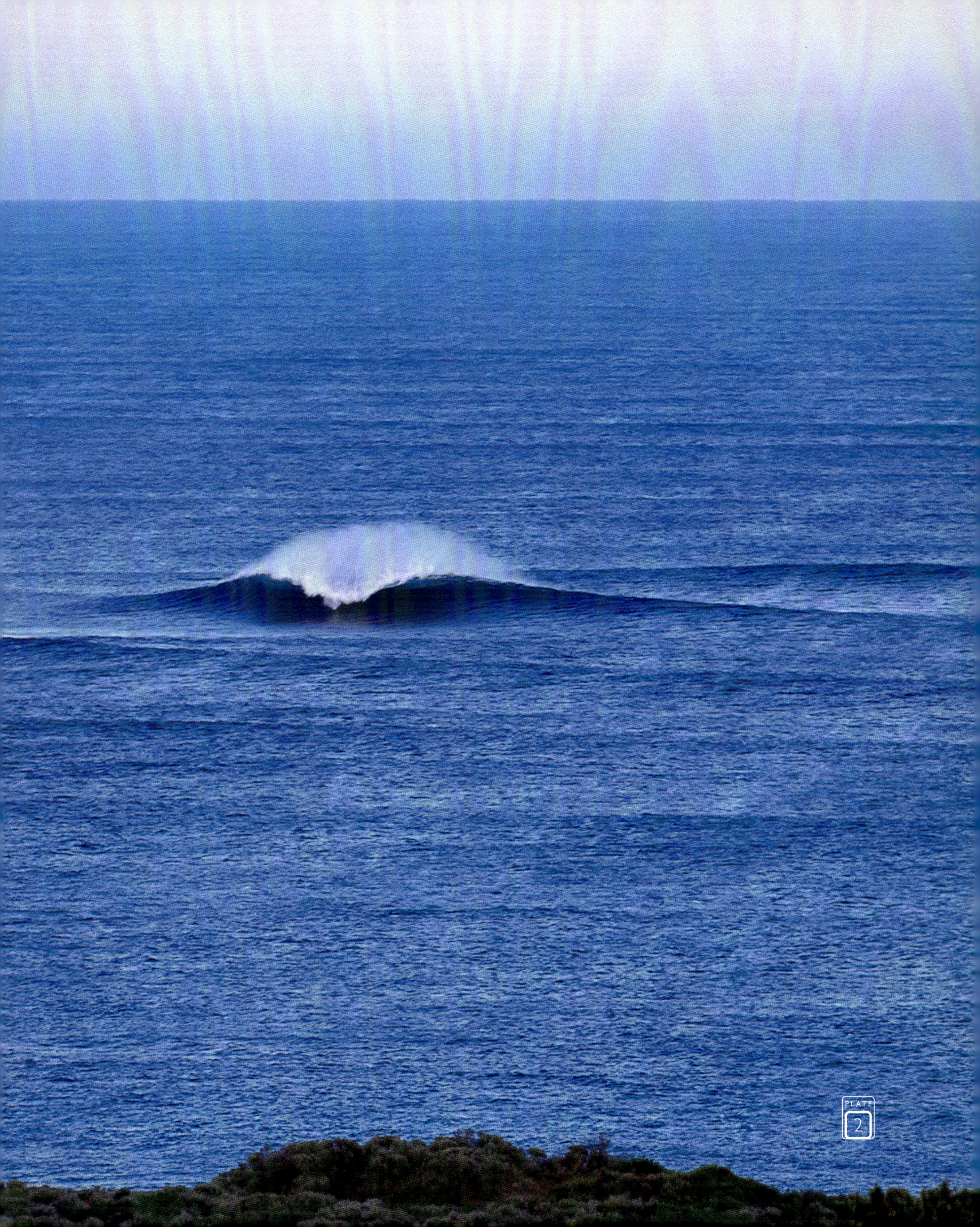

PLATE 2

. . . The waters rose more and more on the earth
so that all the highest mountains
under the whole of Heaven were submerged.

Genesis 7:17-20 THE JERUSALEM BIBLE popular edition

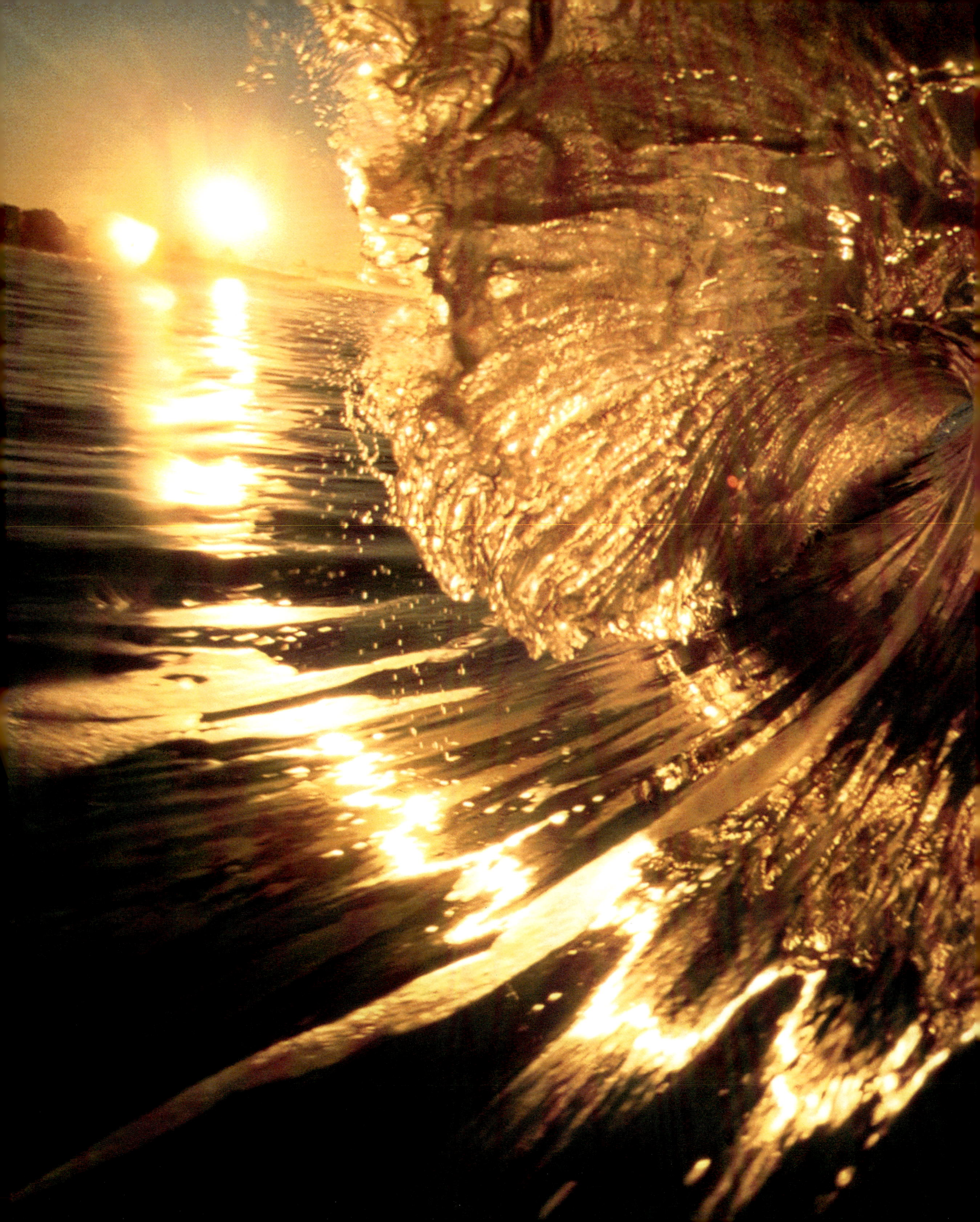

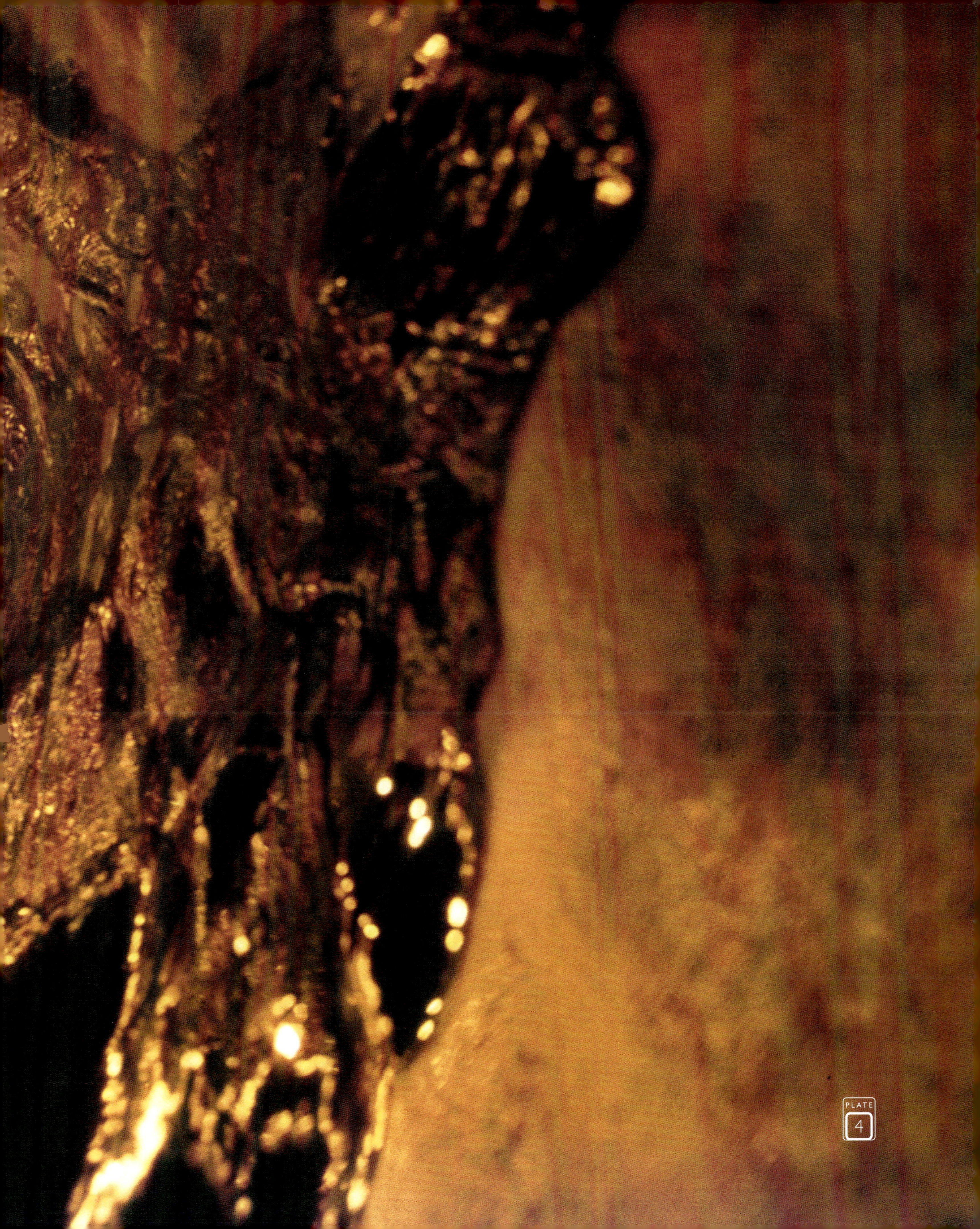

PLATE 4

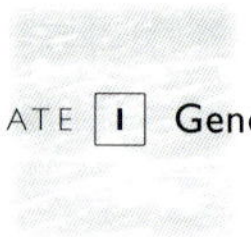

PLATE 1 Genesis indeed. A strong wind whips the ocean off Hawaii, forming the beginning of a sea; given long enough and far enough, the wind's energy will result in evenly spaced swells capable of travelling thousands of miles beyond their source.

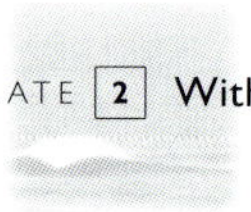

PLATE 2 With lonely symmetry, a peaking swell curls over a reef far offshore in south-west Australia. This coastline, its limestone bedrock exposed to the huge energies of the Roaring Forties and the southern Indian Ocean, is among the wildest in the world.

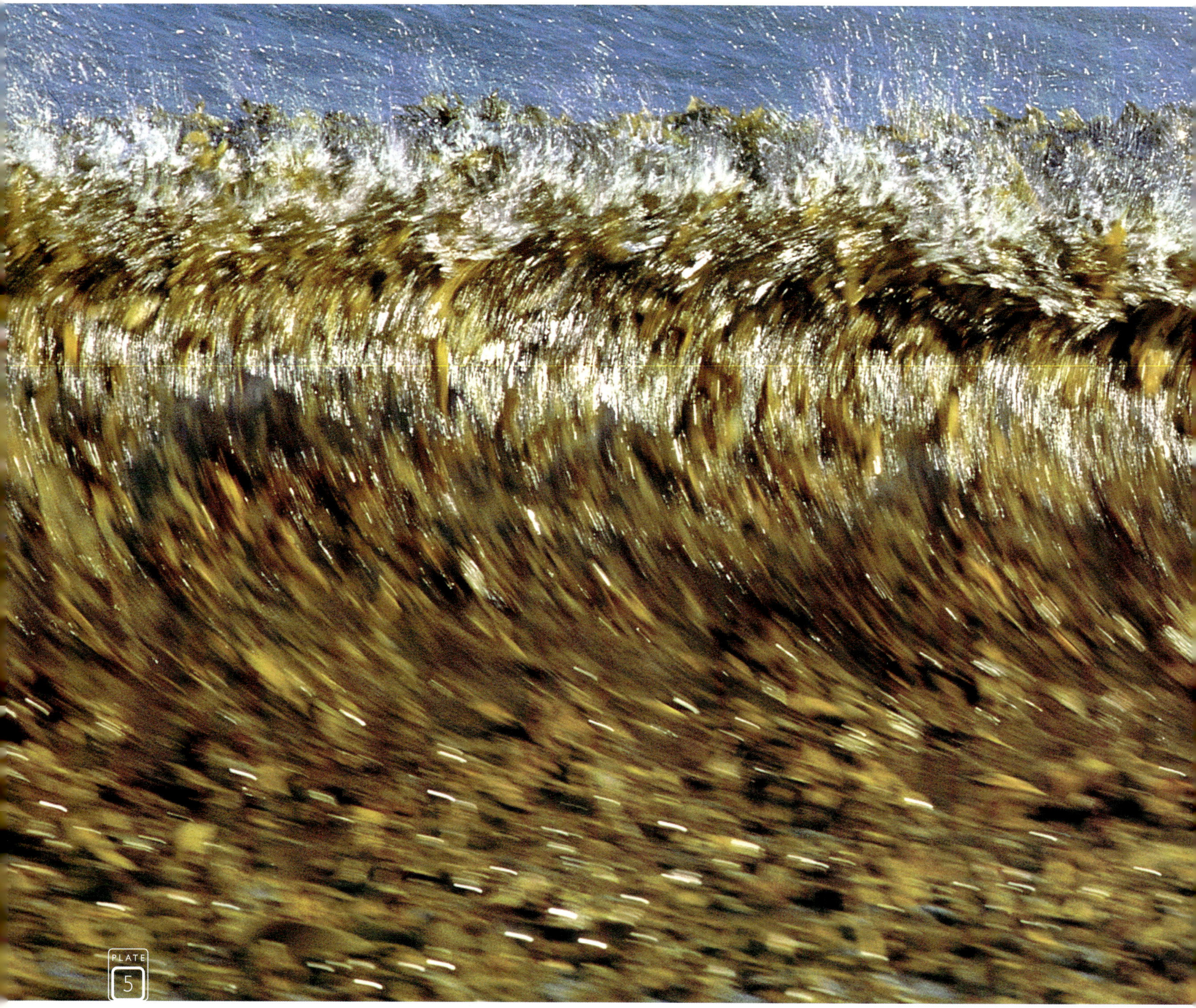

PLATE 5

PLATE 3 Far from its origin in some distant storm, a silent energy pulse lifts Pacific Ocean waters into the familiar curve of a large groundswell. Contrary to their initial appearance, swells move through water, not across its surface, disturbing water molecules in gentle circular motions to extraordinary depths: a long-interval groundswell can shift debris on the ocean bottom at depths of over a thousand feet.

PLATE 4 Eventually, each swell's movements press so strongly against the sea floor that a breaking wave results, with its familiar yet endlessly different arc of plunging lip.

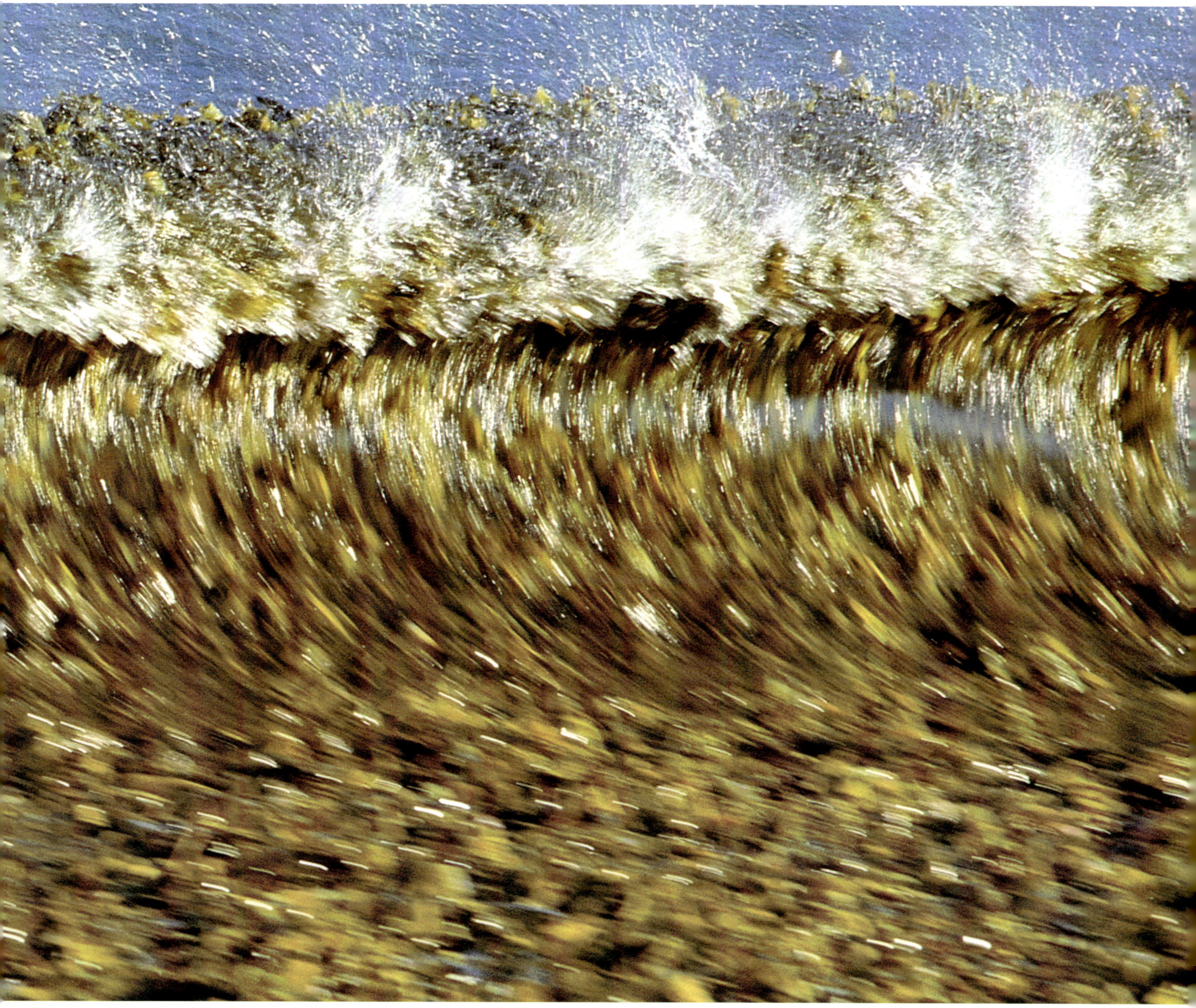

PLATE 5 Elegantly enhanced by a patch of kelp, a small curl pitches forward into shallow water off the east Australian coast. Swells will continue to propagate through such seaweeds with no particular loss of energy. Note the acceleration of water moving up the face of the wave, as it's drawn into the plunging lip.

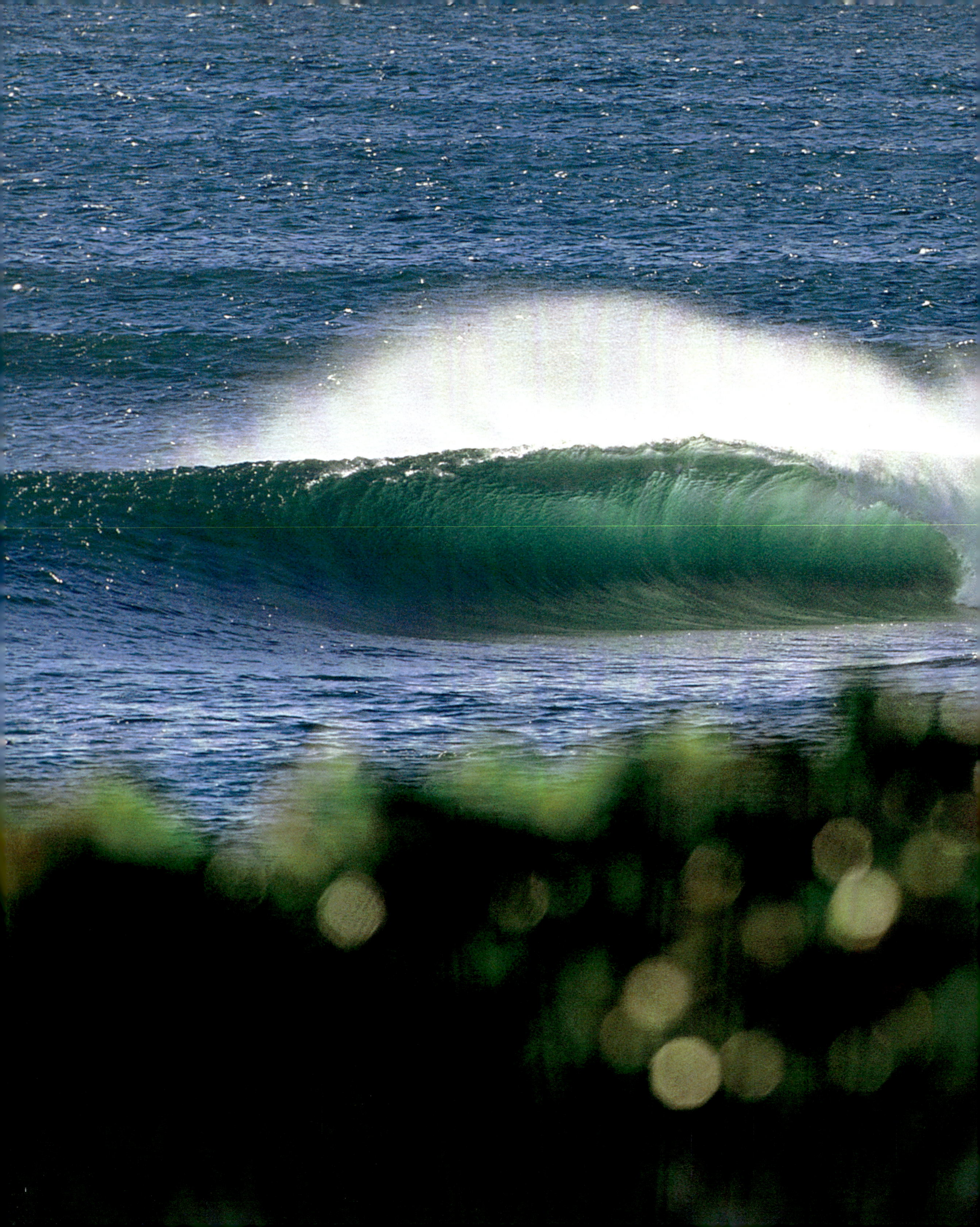

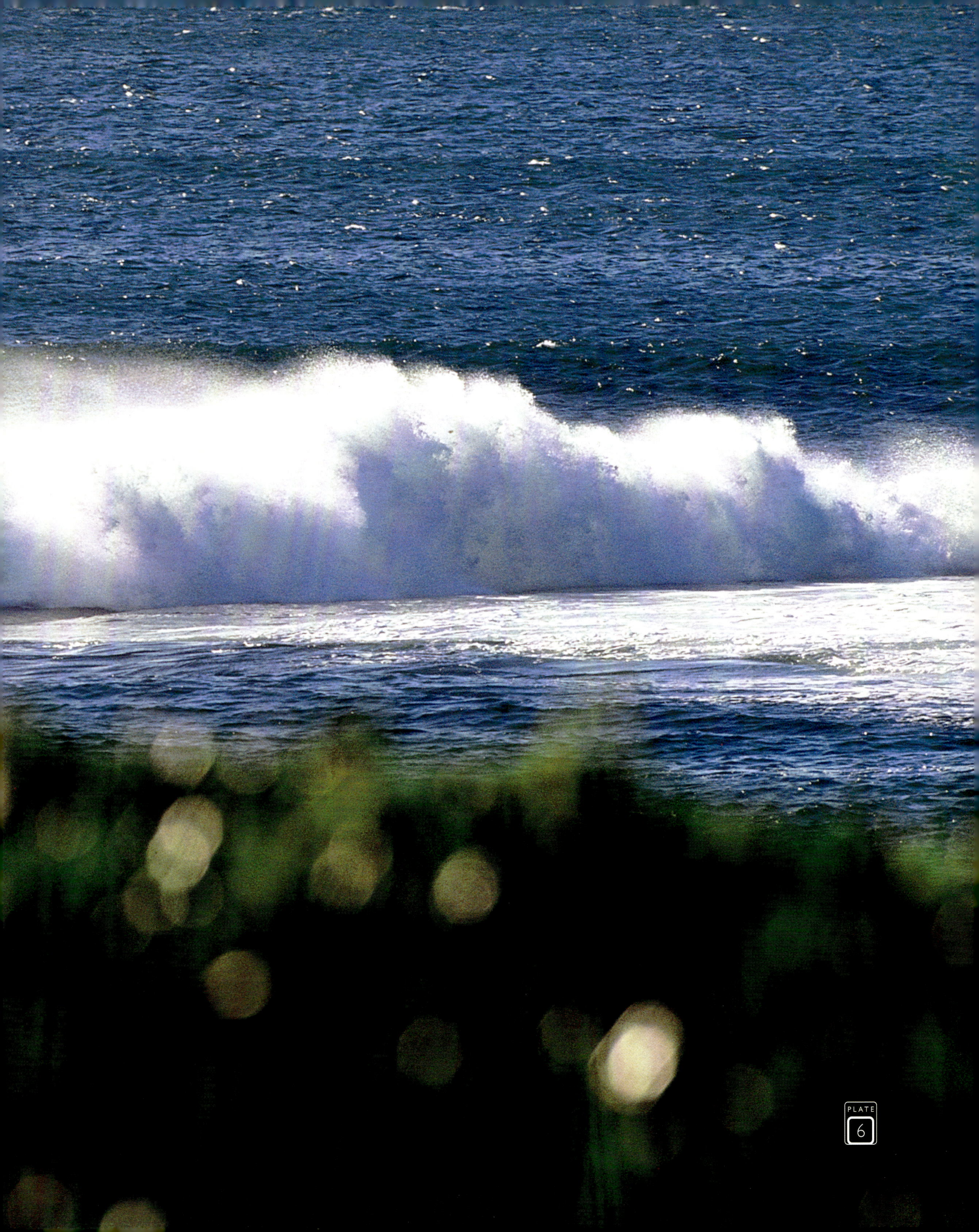

PLATE 6

The three great elemental sounds in nature are the sound of rain, the sound of wind in a primeval wood and the sound of the outer ocean on a beach. I have heard them all and of the three elemental voices, that of ocean is the most awesome, beautiful, and varied.

Henry BESTON THE OUTERMOST HOUSE 1928

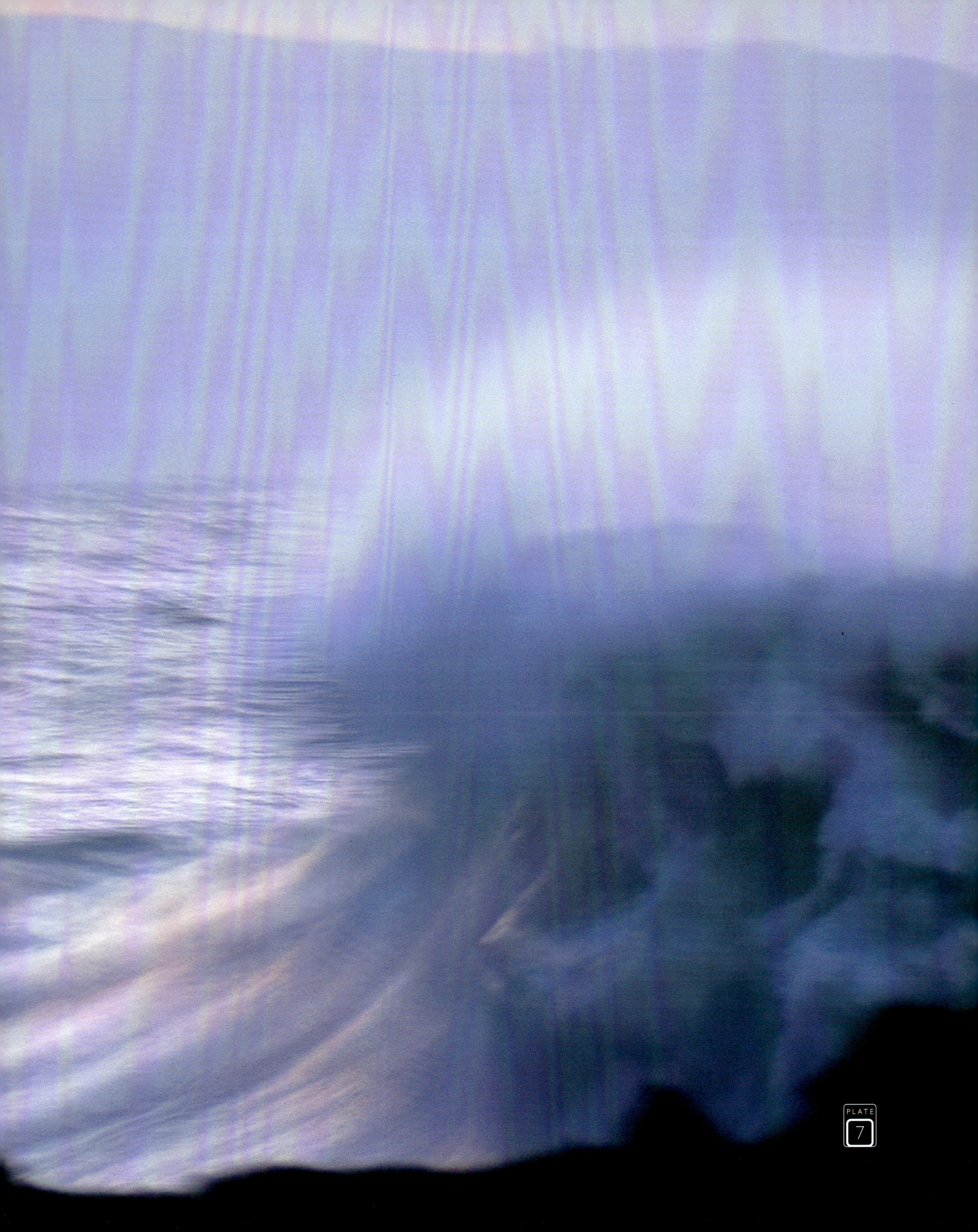

PLATE 7

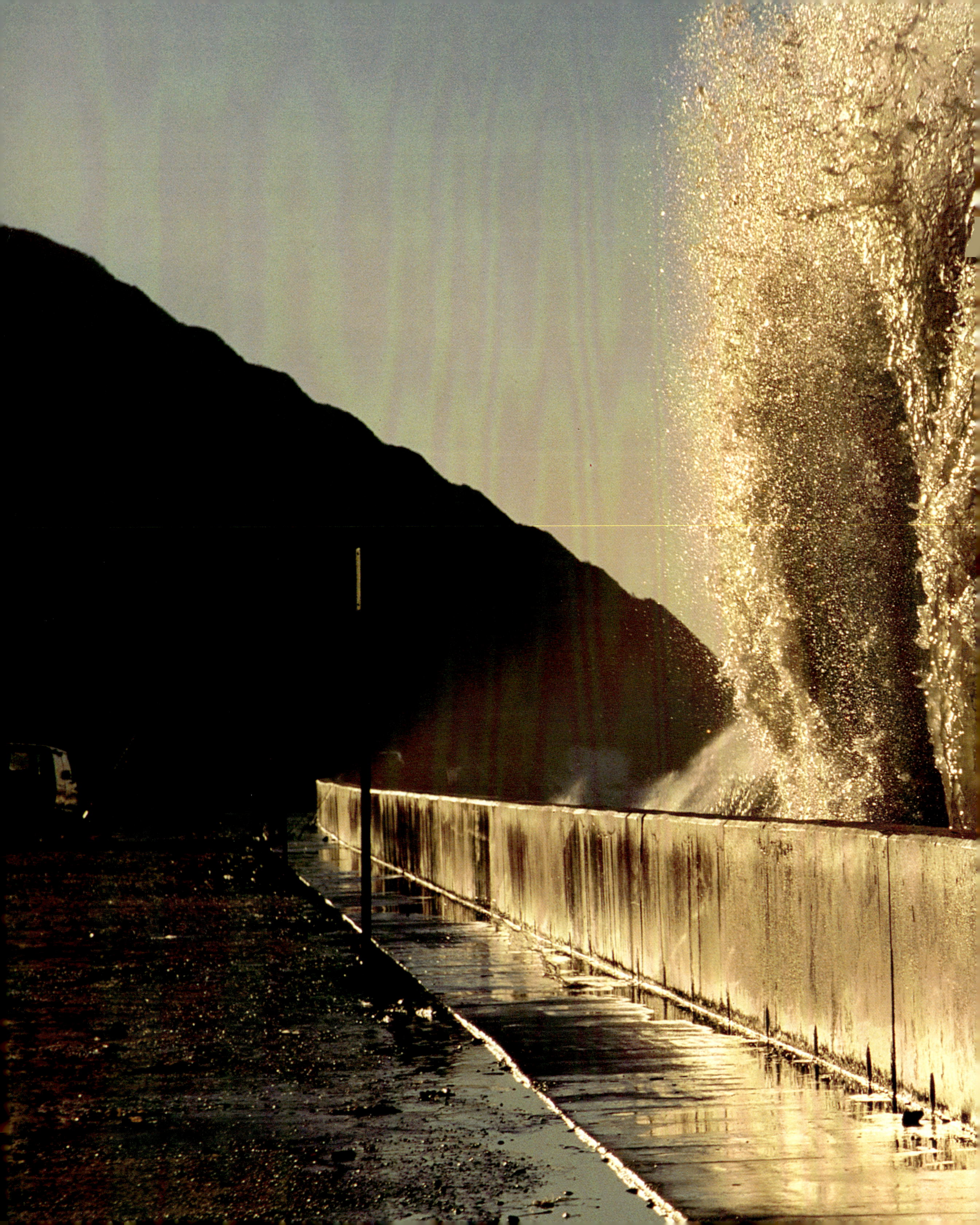

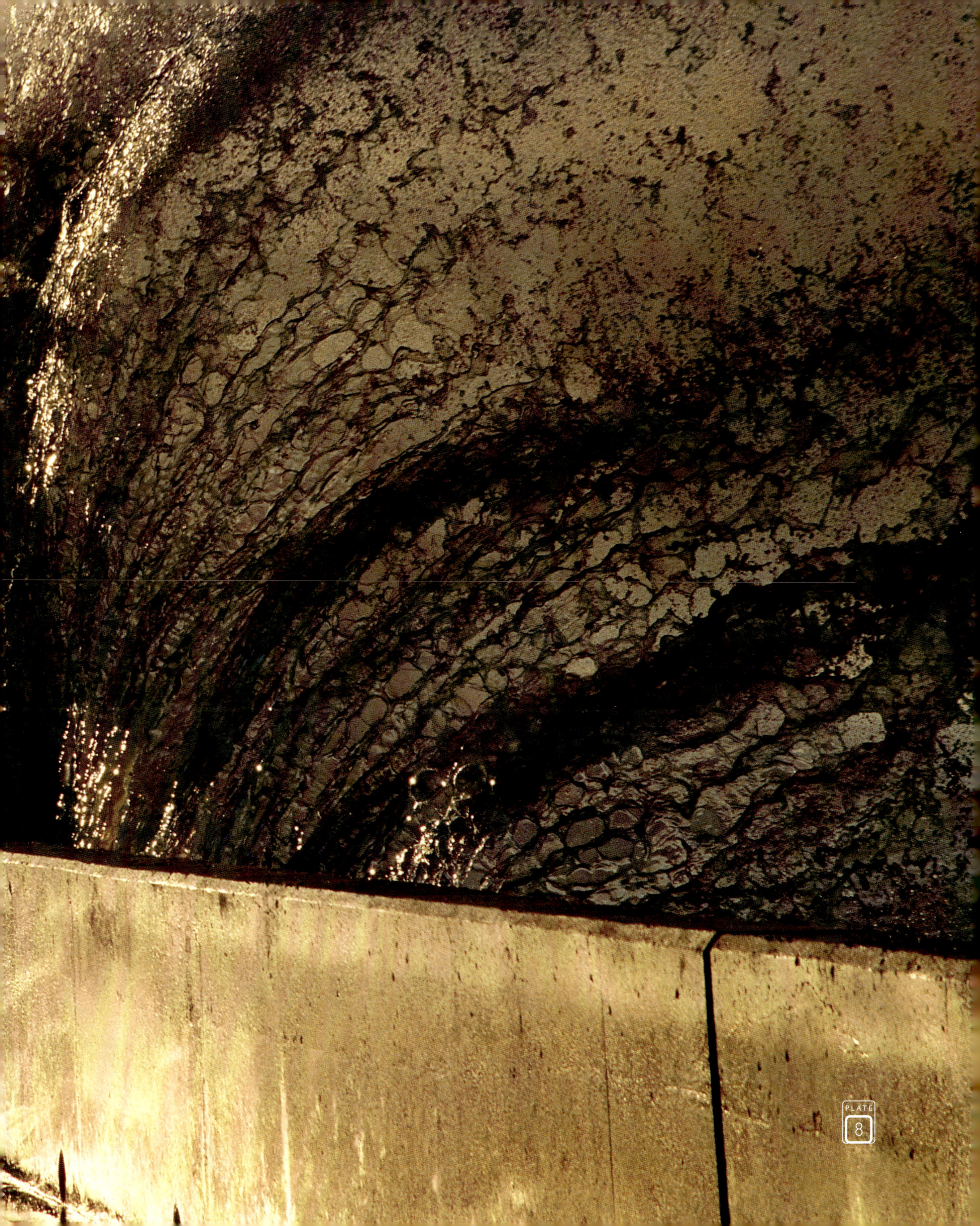
PLATE 8

PLATE 6 Blistering offshore westerly winds greet a fresh groundswell at Australia's Shark Island reef. The offshore wind has little effect on the energy of the long-lined swell, which formed in a storm well over the horizon; instead, it simply sculpts each wave into a clean-faced marvel.

PLATE 7 Here and gone in a blink, waves along a coast can seem ghostlike, yet their effects are very real — as countless eroded beaches and pitted rock platforms attest.

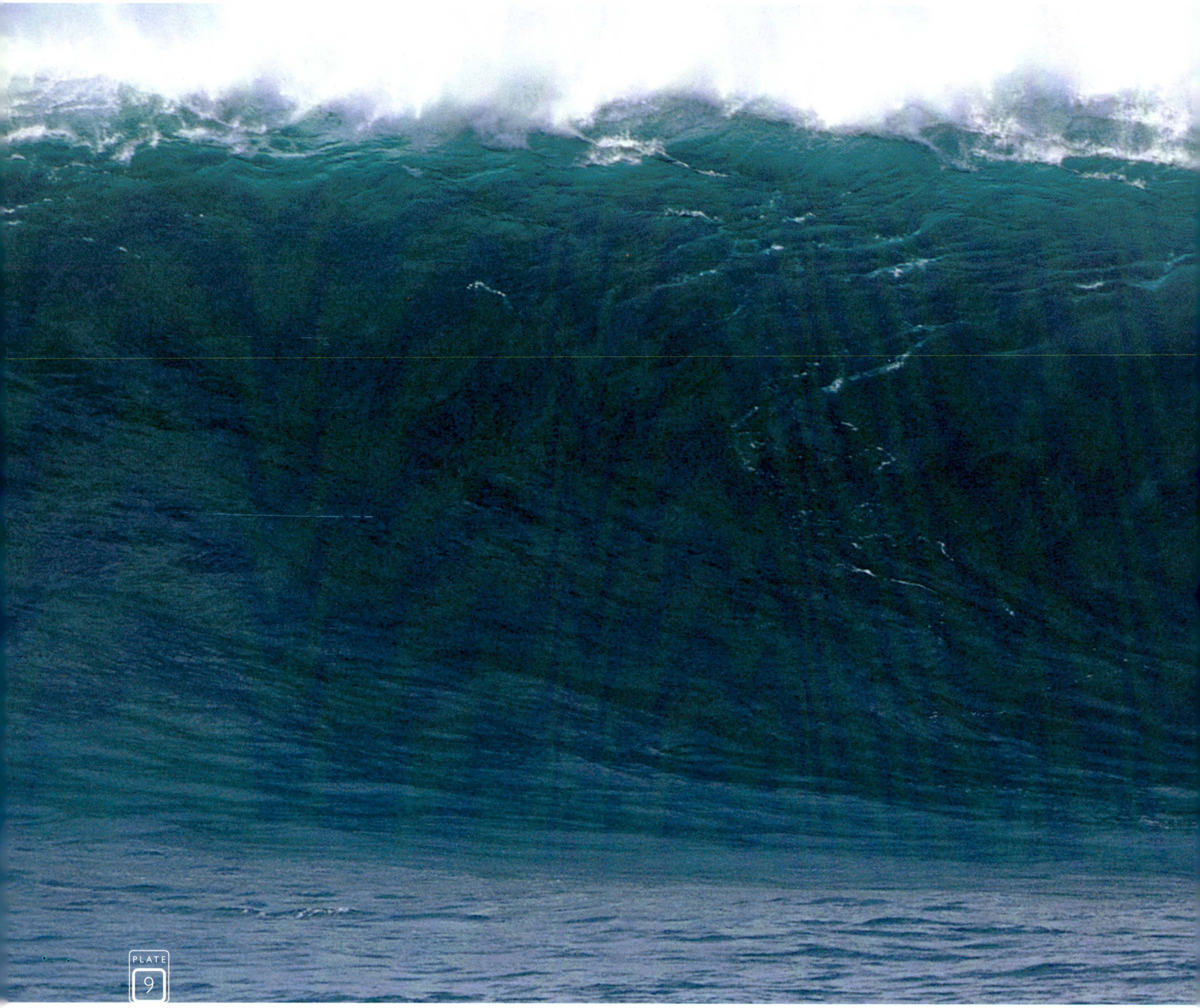

PLATE 9

PLATE **8** El Niño-inspired high surf crashes against a seawall in Ventura, California. The well-known climatic phenomenon, named generations ago by Peruvian fishermen whose lives change dramatically upon its arrival, periodically drives warm water across the Pacific, stunting fish populations and causing northern storms to track lower into the U.S. west coast. The result? Big surf for the waveriders, and panic among coastal real estate owners.

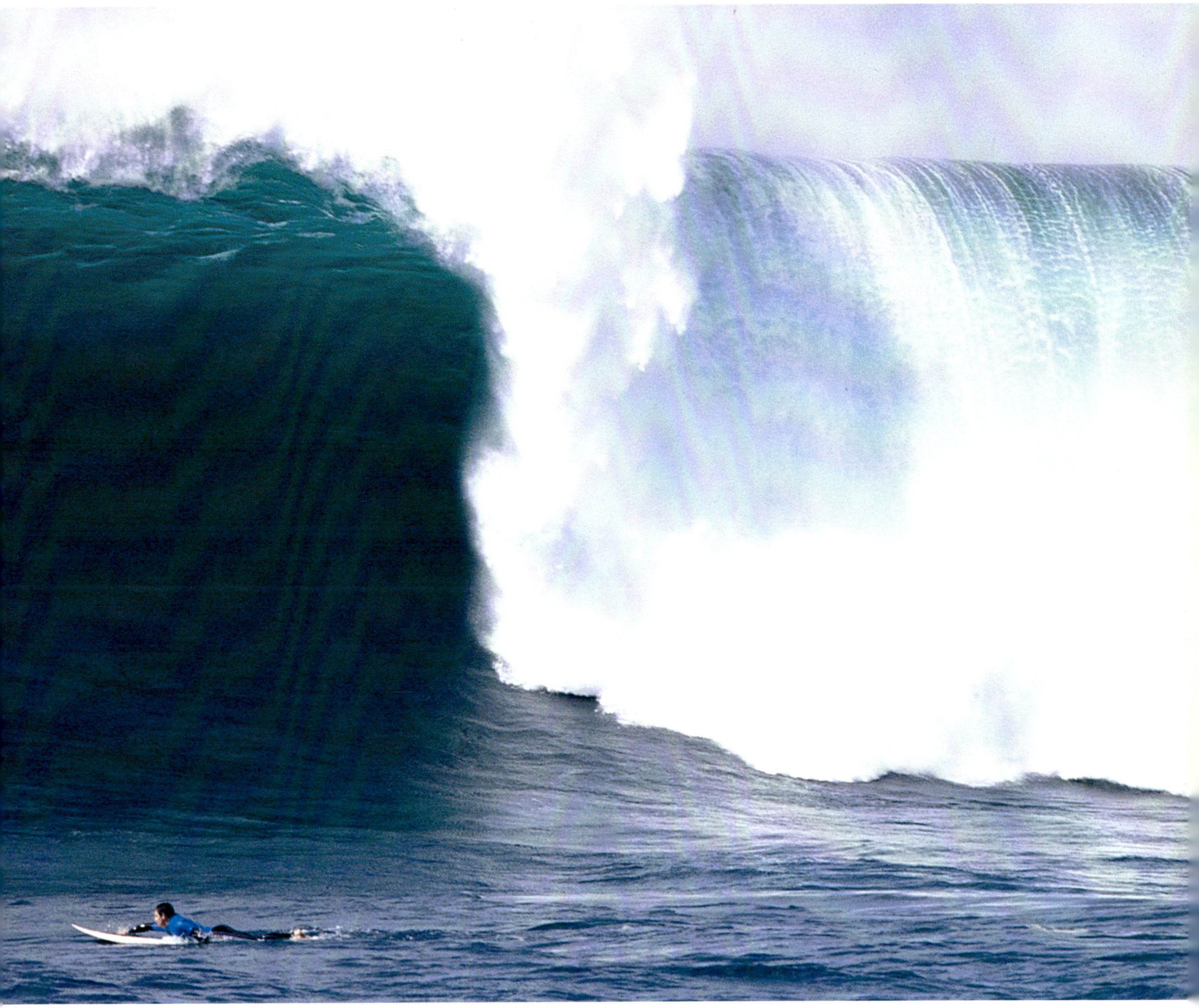

PLATE **9** When encountered further offshore, such swells can send shivers down the most experienced of wave-riders. Terrence McNulty views a thirty-footer with some trepidation at Todos Santos Islands, Mexico.

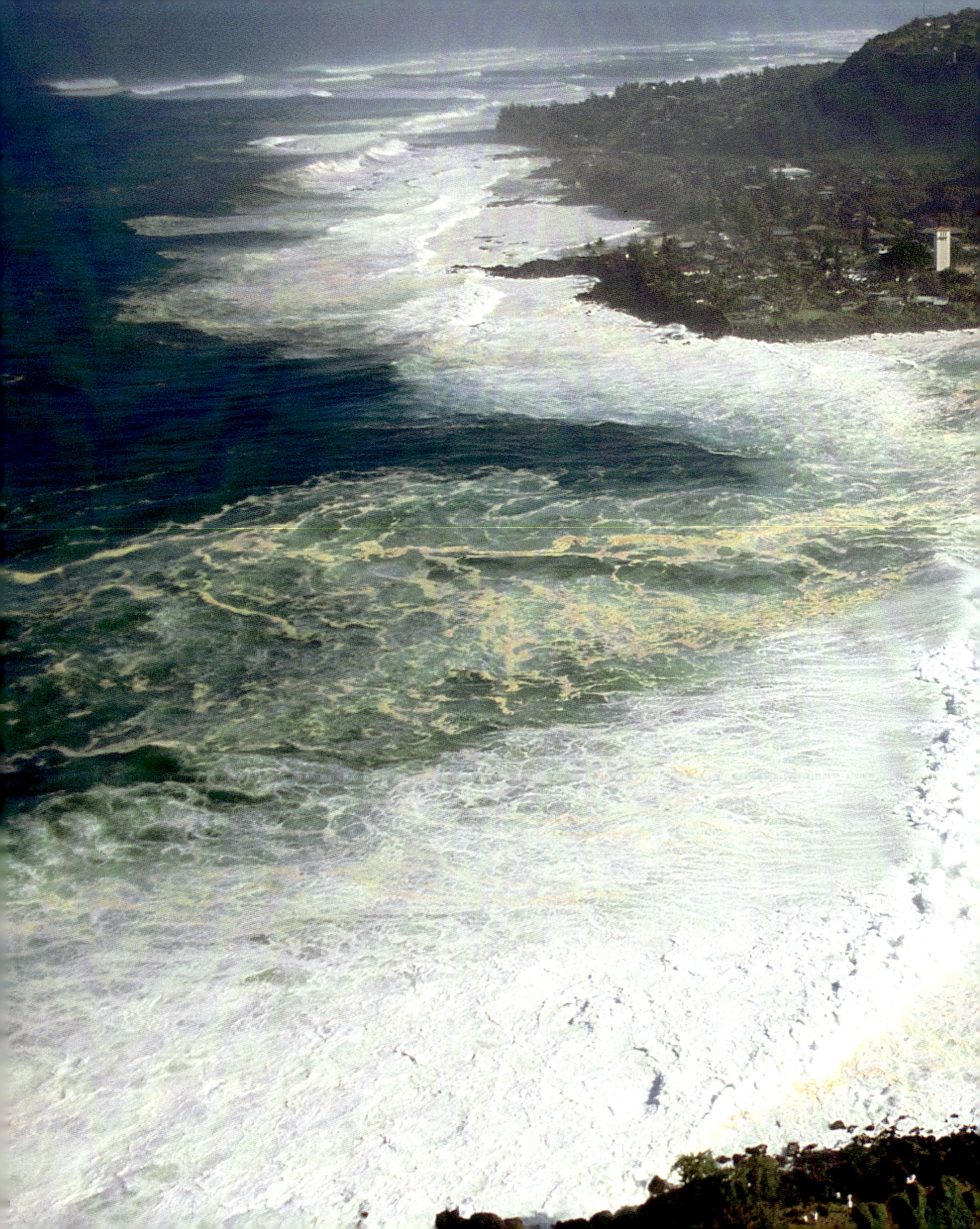

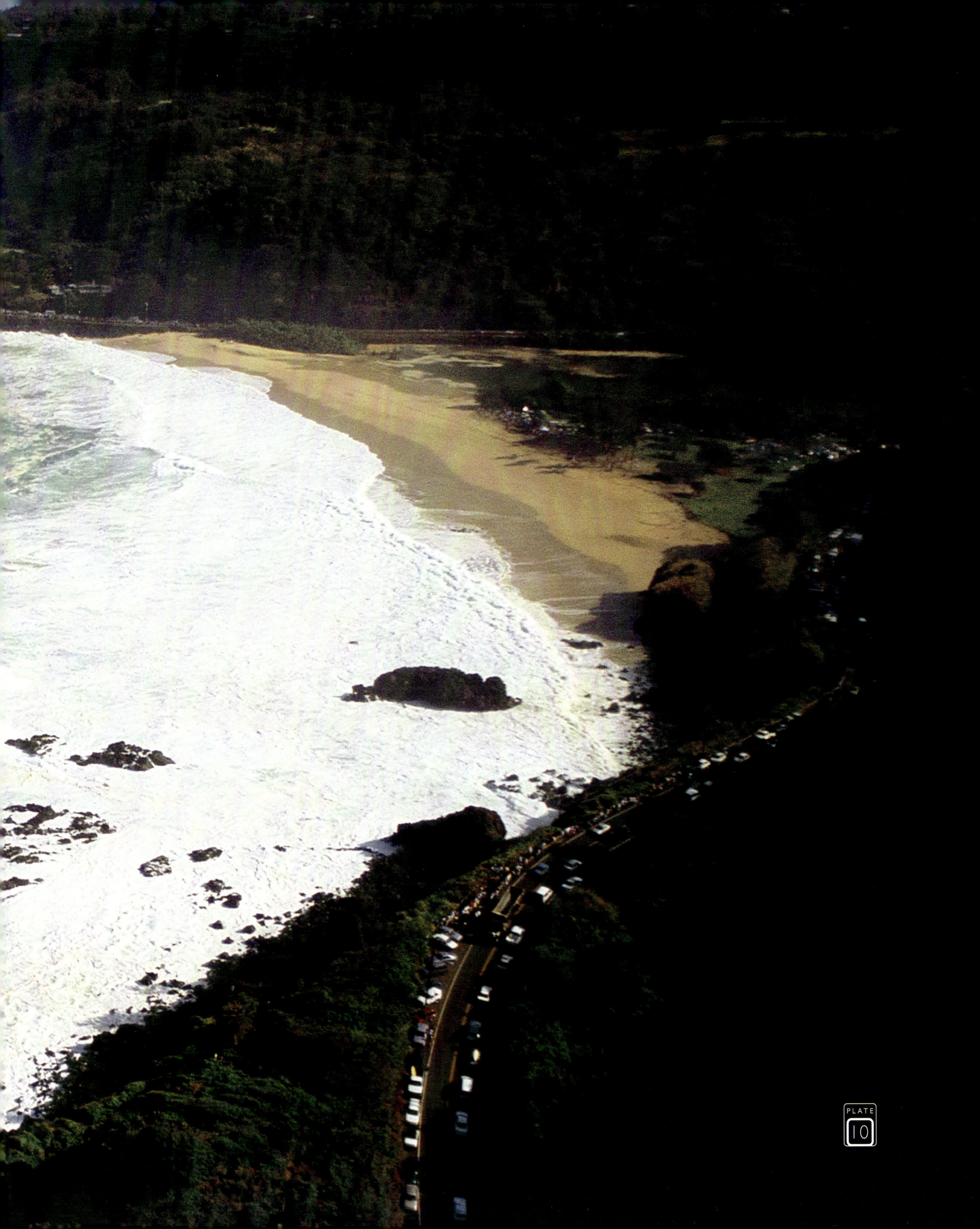

PLATE
10

PLATE 10 El Niño's effects on surf along the North Shores of the Hawaiian chain are famous among surfers. This aerial image shows Oahu's Waimea Bay under full assault by a thirty-five-foot-plus north-west groundswell on January 28, 1998. This legendary day saw only a couple of surfers attempt Waimea; neither successfully rode a wave. Note the tremendous rip drawing the influx of swell and foam back out through the middle of the bay. Also worth noting, high in the top left corner of the shot, are foam lines on the Outer Log Cabins reef, where earlier in the day Hawaii's Ken Bradshaw was towed into the biggest wave yet ridden by humans.

PLATE 11 On calmer days, Oahu's Pipeline breathes oceanic fire just north of Waimea. It's said that in any photograph of an exploding Pipeline wave, you'll find a face somewhere in all that foam and spray — nobody's yet decided whether it's the face of God, or the Devil.

PLATE 12 A complex interplay of forces can be seen clearly from within as this small wave plunges onto a California beach. Streaks of falling water in the lip are latticed with ripples created when the wave first began to elevate across a sandbar, all lit by the refracted early morning sun. A split second later — and it's gone.

PLATE 13 Peeling between sea and sky, a swell slithers along the Indian Ocean coast north of Perth, West Australia. Coastlines, though seemingly permanent, are anything but; just a few millennia ago during the last Ice Age, when much more of the ocean's waters were stored in glaciers and icecaps, this wave would have broken on the exposed edge of the continental shelf, some ten kilometres out.

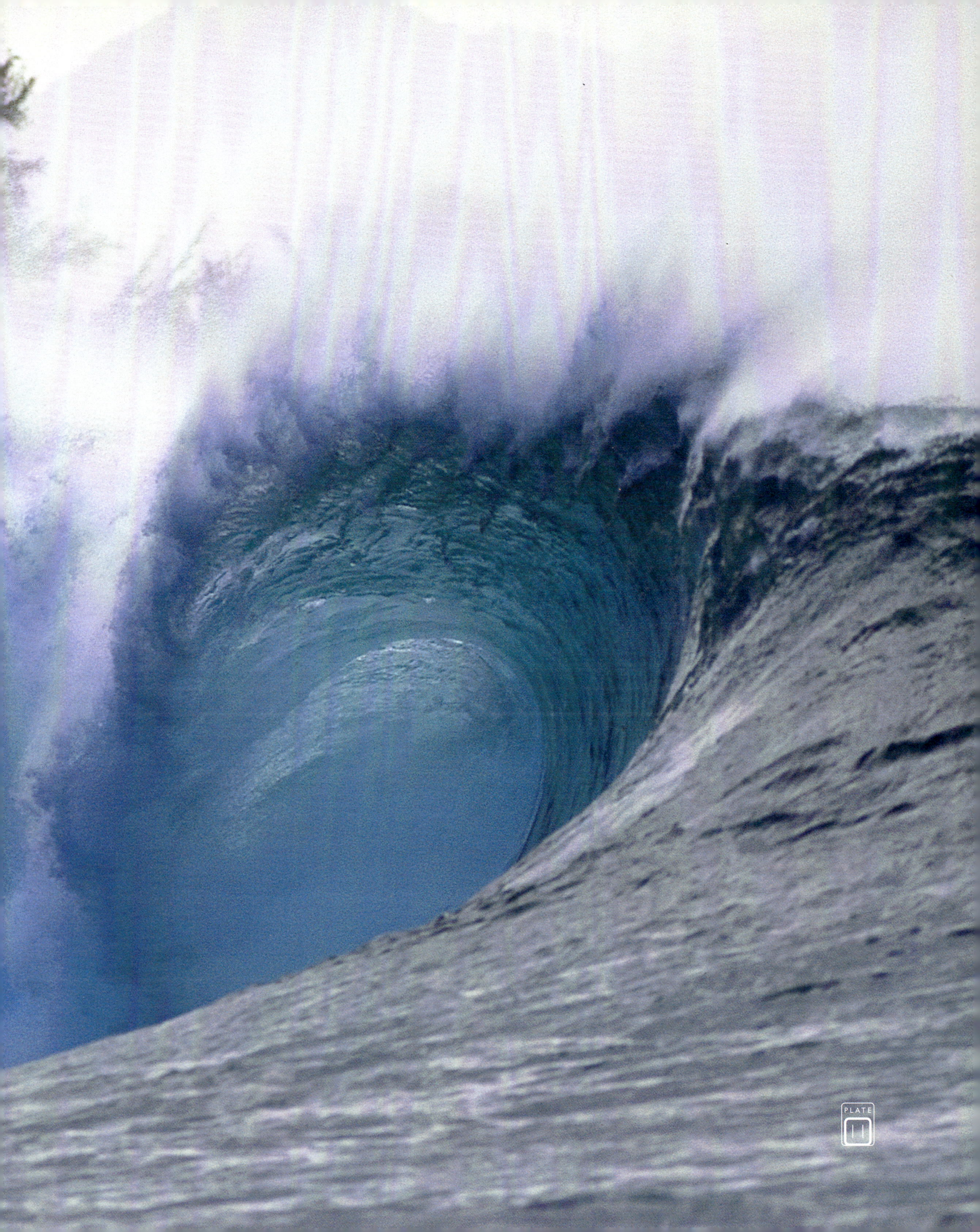

PLATE 11

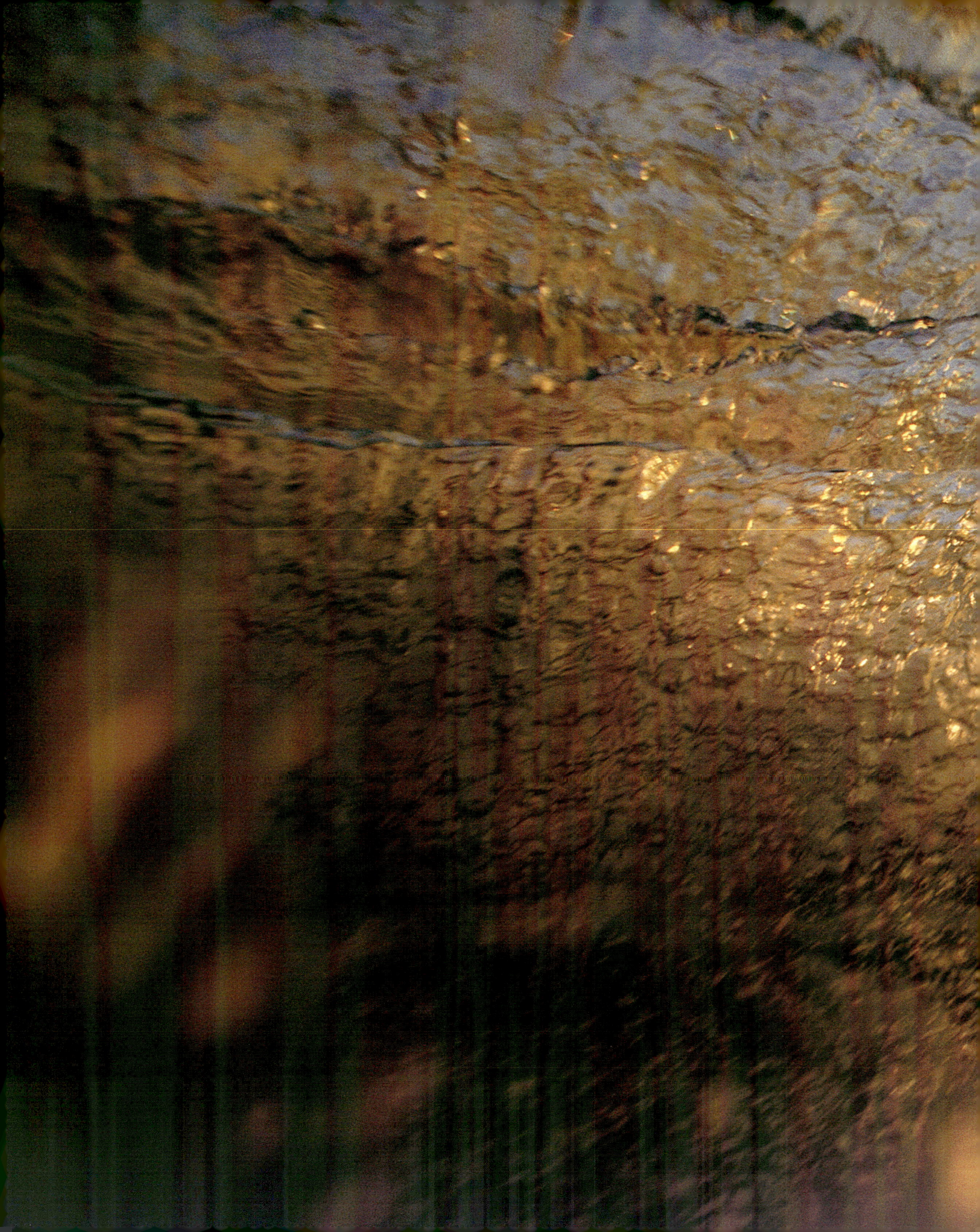

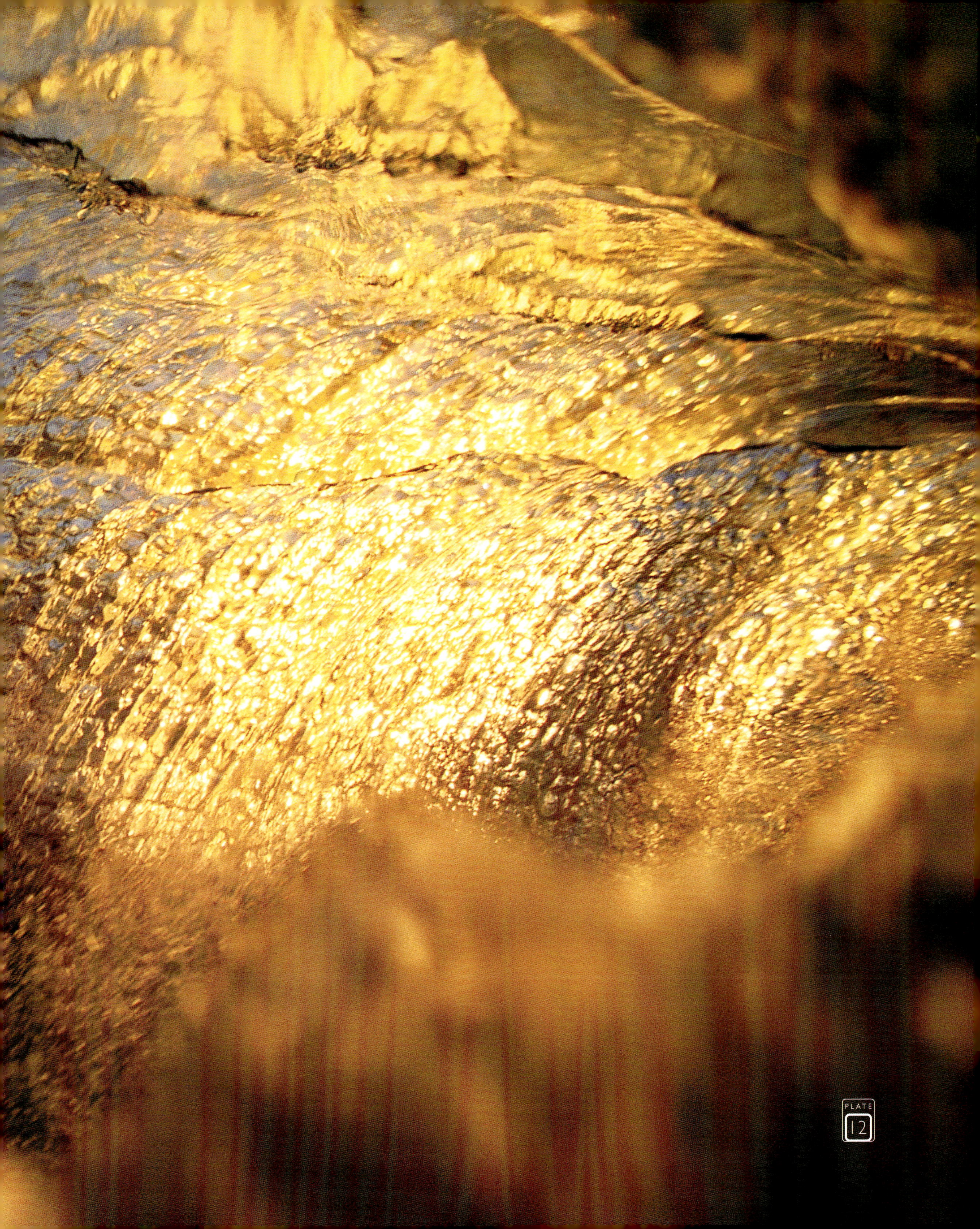

PLATE
12

If the waves had their way about it – with no interference from the upthrusting or downwarping of the coastline – all the beaches in the world would be perfectly straight, and slope imperceptibly seaward…

. . . so that the waves would break very far out and die away completely before reaching the shore. This is what is called a mature shoreline, and represents the ultimate in beach stability. Although the waters may sweep inland many miles during severe storms, they tend later to drain off through local stream outlets, leaving the beaches relatively unaffected. Mature shorelines are characteristic of the southeastern Atlantic coastline of the United States and most of the Gulf of Mexico. In these areas, one can walk half a mile to sea without going over his knees, but he cannot get hurricane insurance unless he lives more than 50 miles inland.

In contrast, the western coasts of America are zones of active mountain building, and thus have few mature beaches. Although the waves are steadily chopping away at the headlands—upon which their energy tends to be concentrated by the process of refraction—they have not yet succeeded in their eventual task of straightening out the coastline and wearing away the cliffs. Here the products of cliff erosion are accumulated in successive concavities (pocket beaches) along the shoreline. Such beaches are ephemeral and precarious, and the sand can be rapidly swept away during a period of high waves, leaving the cobbles exposed. It is really astonishing how rapidly a major increase in wave intensity can cut back a stable, gently sloping beach. I have seen storm surf lower the sand level by 15 feet overnight off mile-long beaches in La Jolla. Within a few hours, a previously safe beach becomes an extremely dangerous place for the inexpert swimmer.

Although the sand tends to be returned to shore during calmer periods, the ultimate action of the waves is to cast it above the tide limit during storms—where it can later be carried inland by prevailing winds—or to move it along the shore until it encounters a submerged channel or submarine canyon, down which it flows to depths immune to wave action. Thus immature beaches are continually being destroyed and renewed by waves, at the gradual expense of the cliffs behind them, and it is only a question of time before the sea will win the battle—but not for a few hundred million years!

William G. Van DORN OCEANOGRAPHY AND SEAMANSHIP 1974

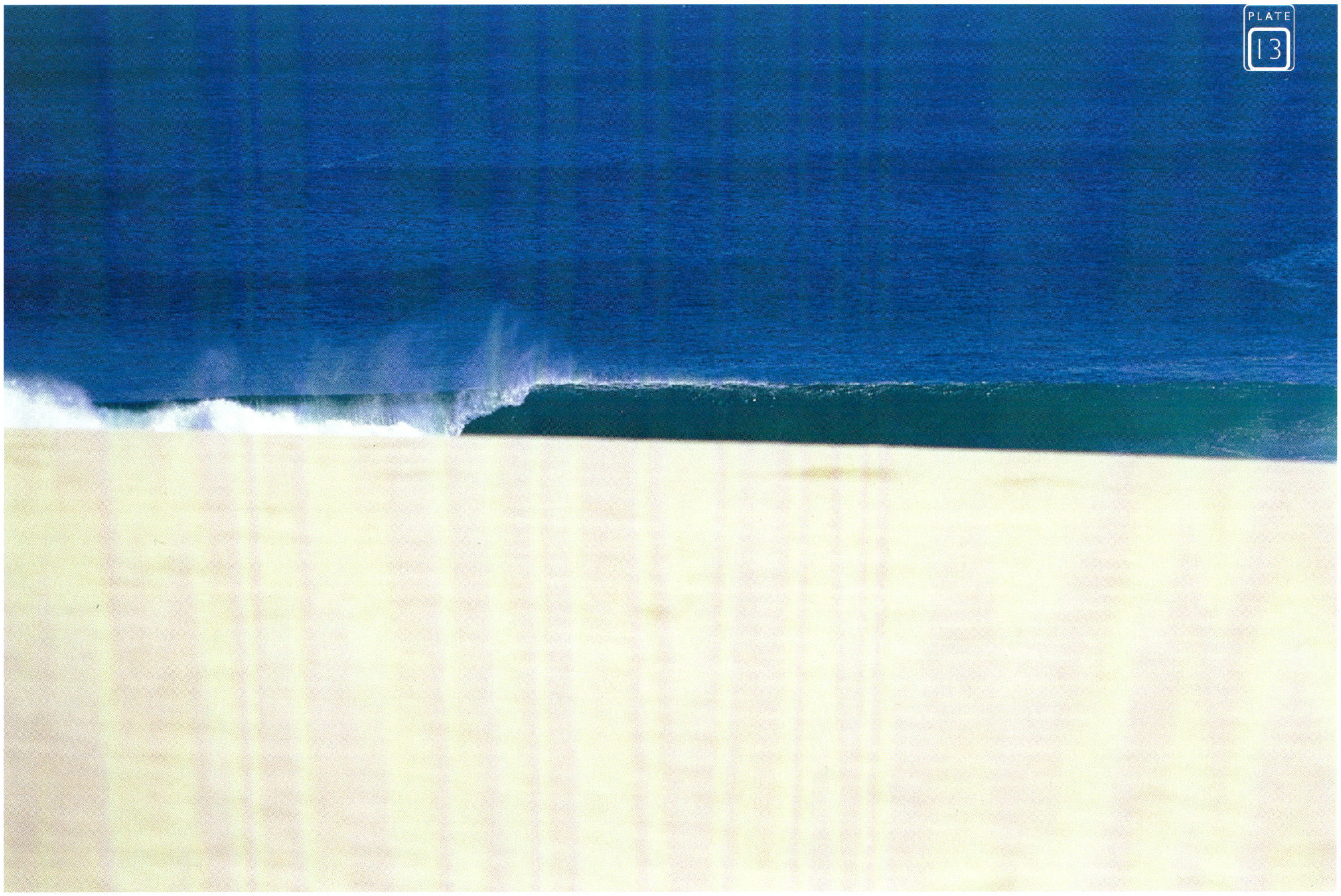
PLATE
13

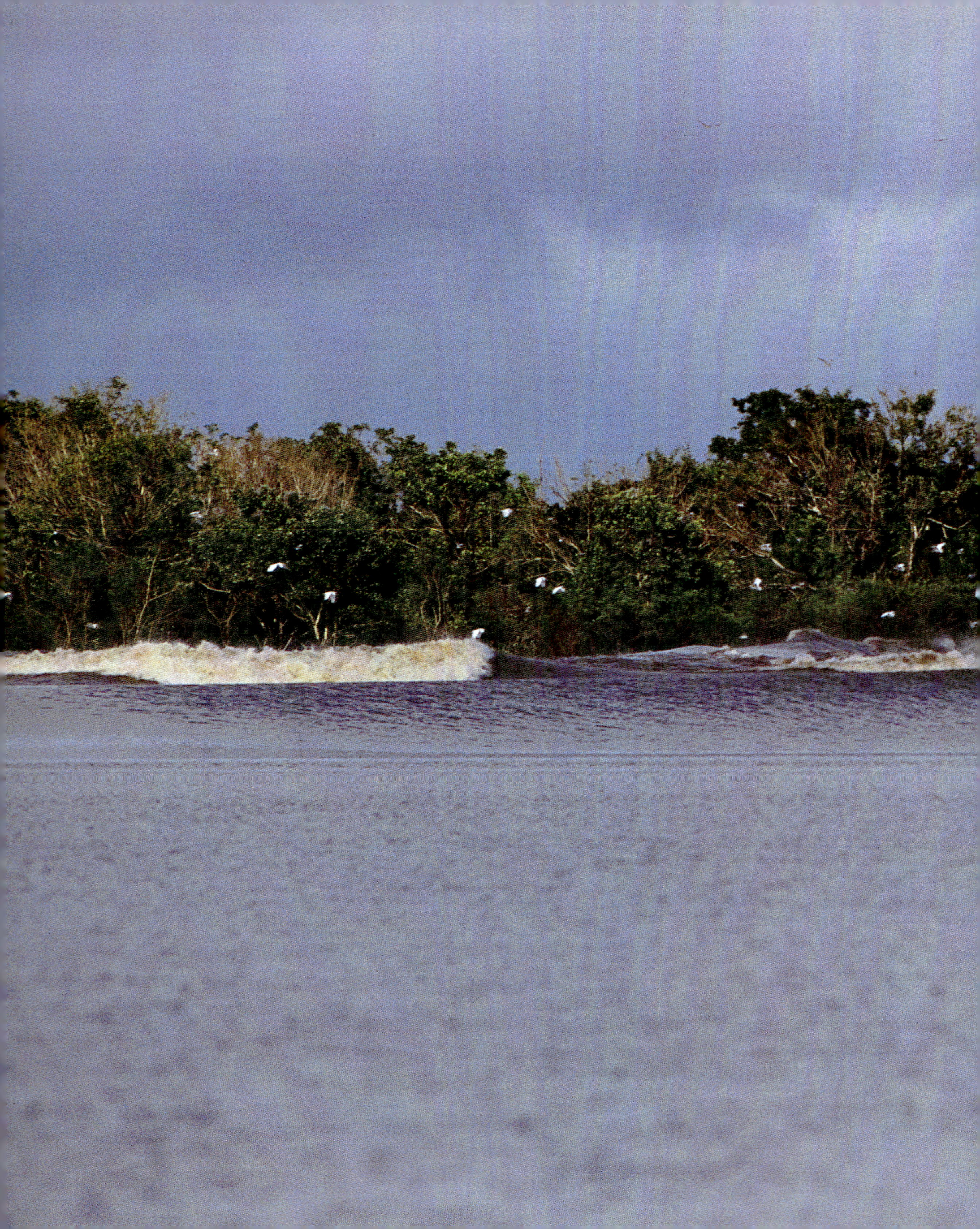

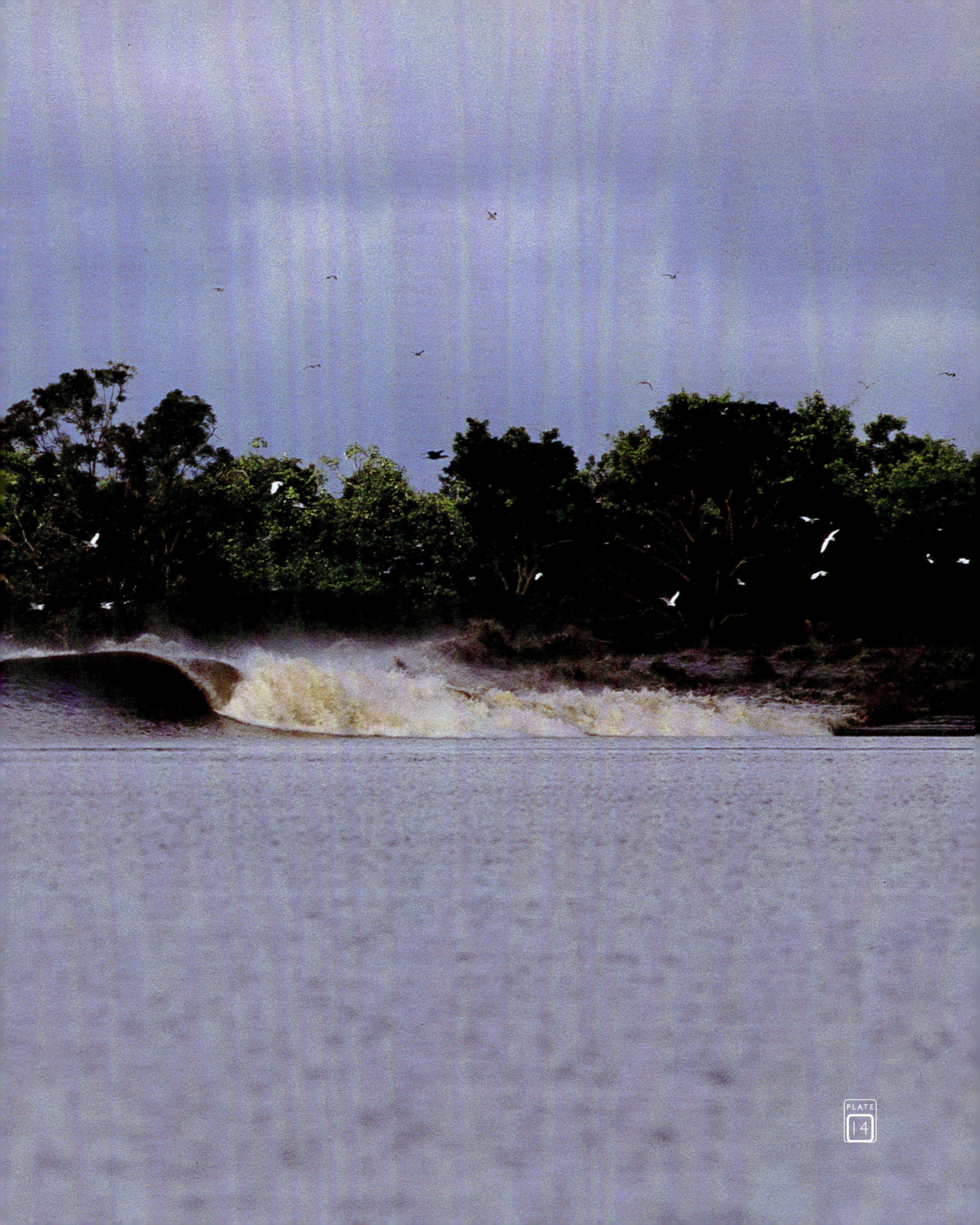

PLATE 14

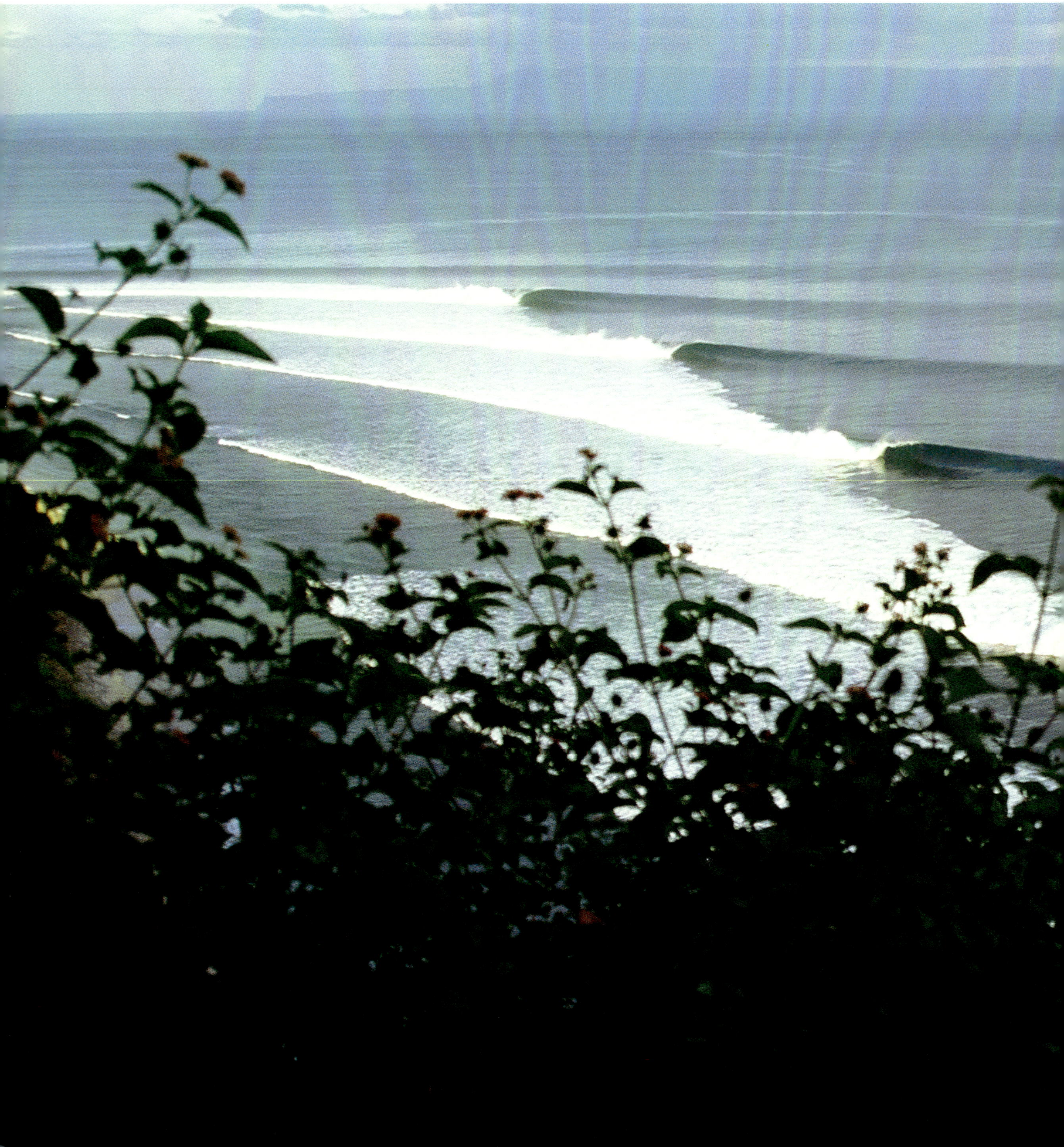

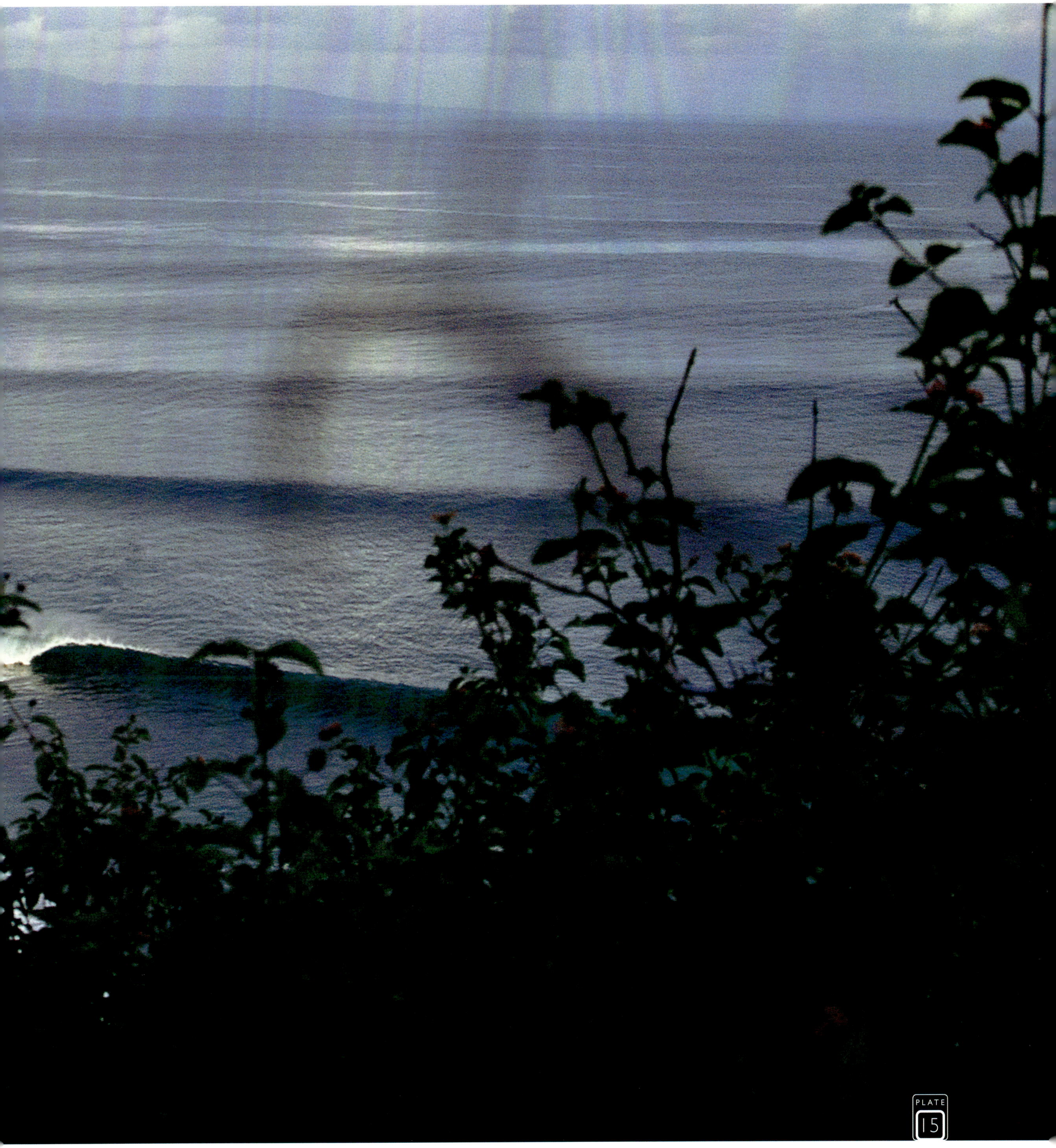

PLATE 15

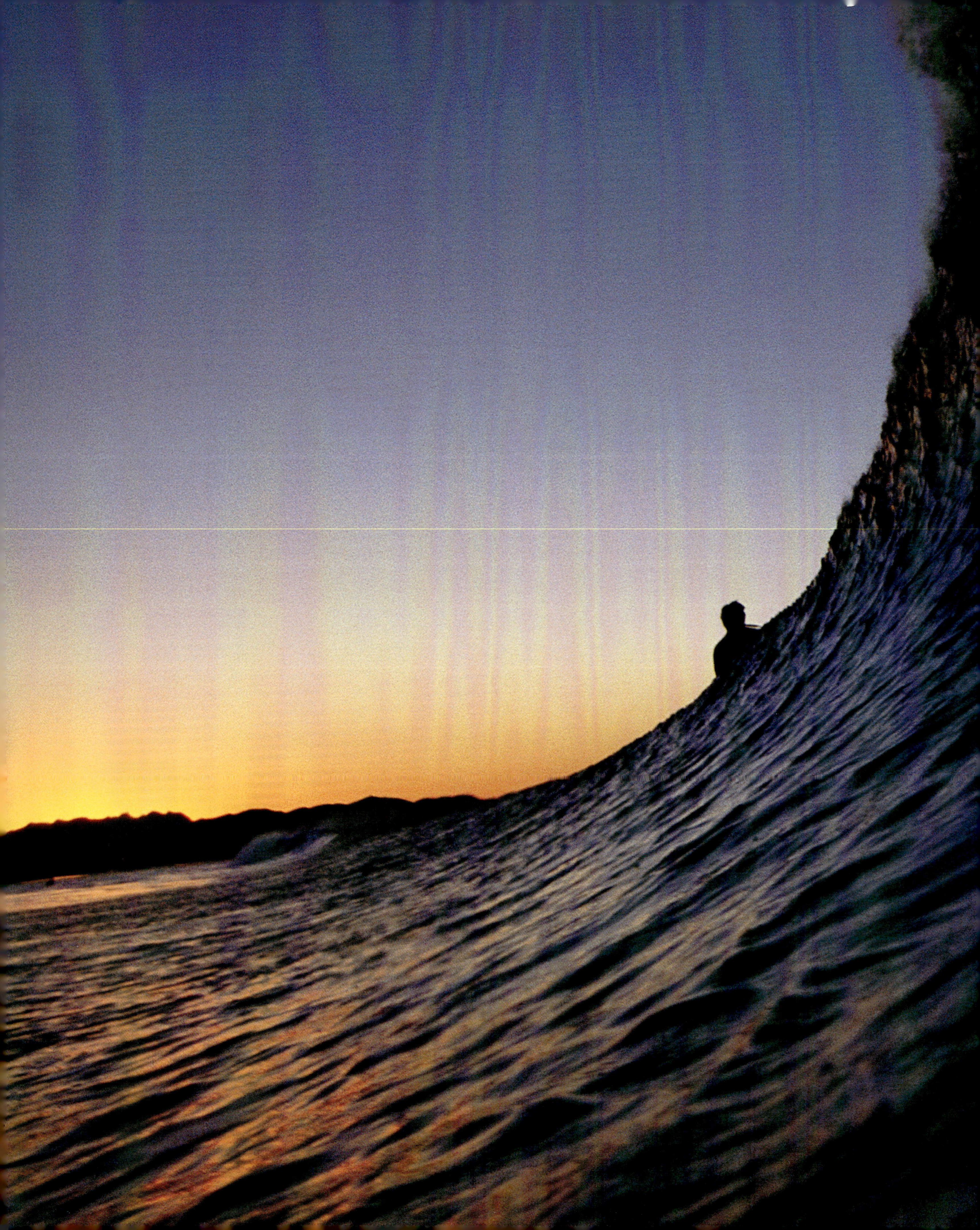

'Broken Buckets' was our nickname for the waves at Broken Head, and if you were to rest a bucket on its side you would get an idea of the shape of the break. They were unique waves, very difficult to see coming and therefore hard to catch. We had to paddle our hearts out for the slightest movement ...

PLATE 16

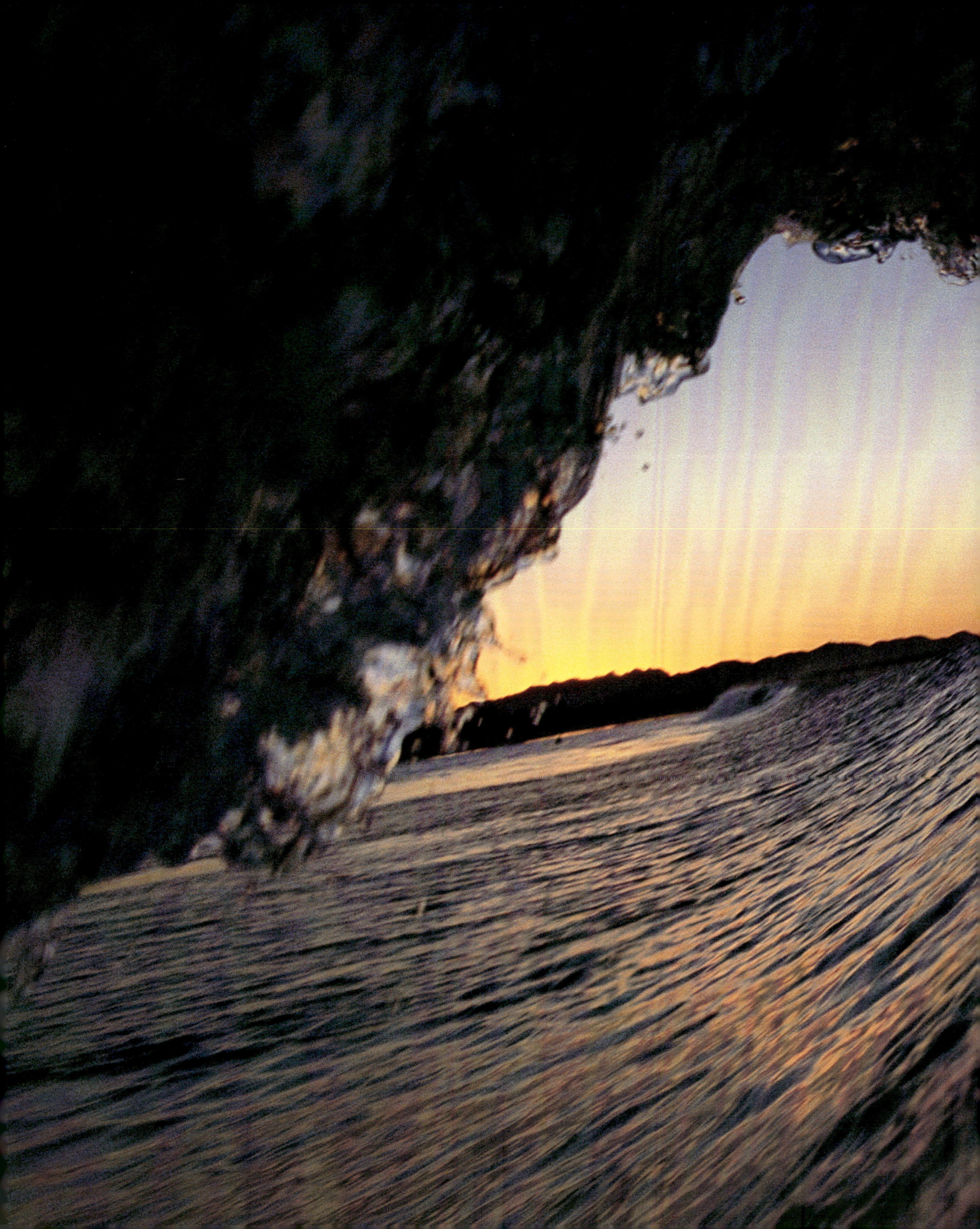

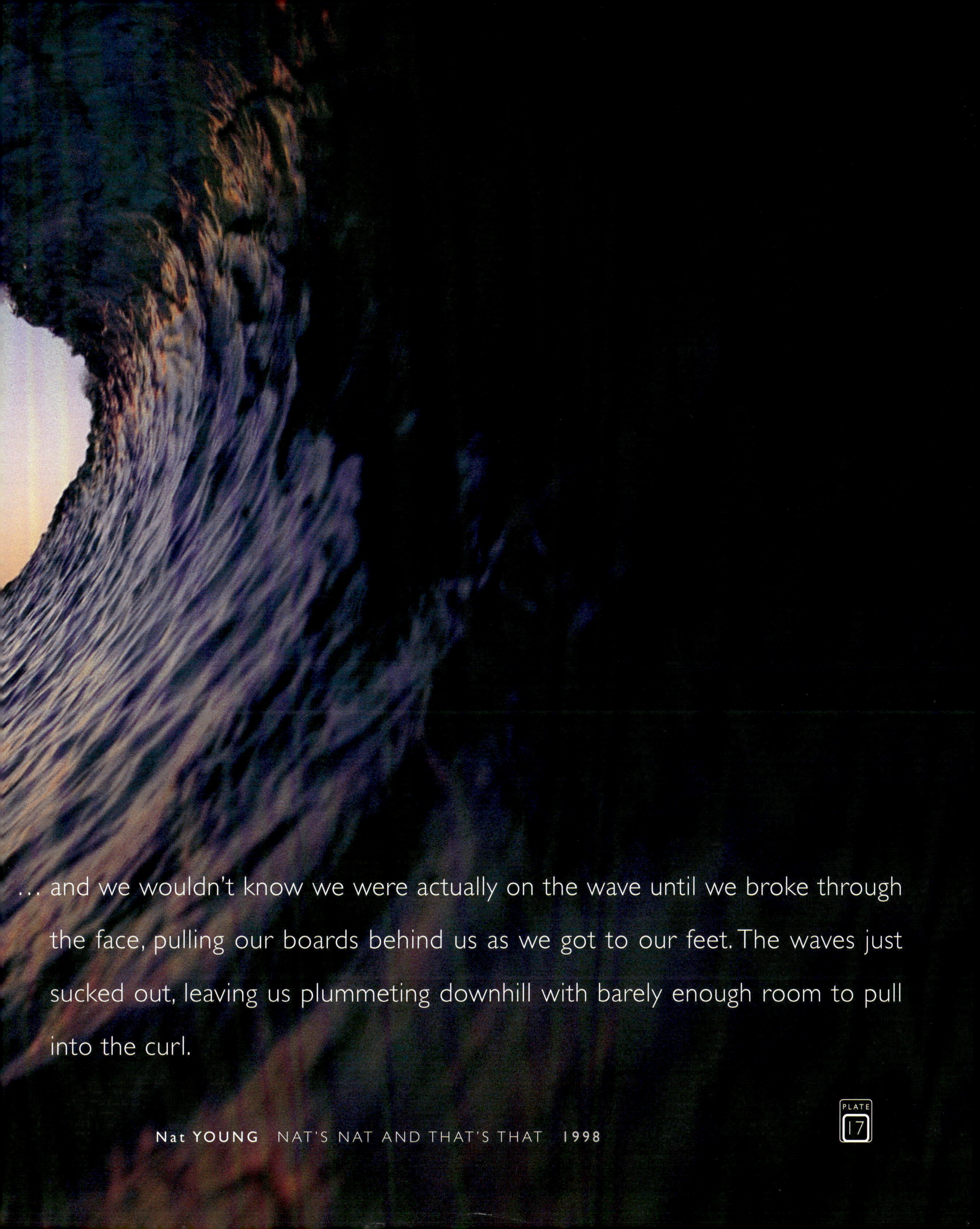

… and we wouldn't know we were actually on the wave until we broke through the face, pulling our boards behind us as we got to our feet. The waves just sucked out, leaving us plummeting downhill with barely enough room to pull into the curl.

Nat YOUNG NAT'S NAT AND THAT'S THAT 1998

PLATE 17

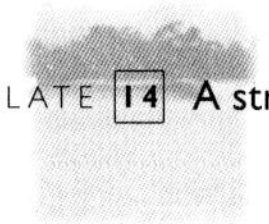
PLATE 14 A strange vision of the coast travels upstream in the fabled Amazon River, Brazil, scaring the wildlife in the process. This tidal bore wave, caused by the pressure of a rising ocean tide against the river's seaward flow, is a regular visitor to the lower reaches of the Amazon, and has been ridden by surfers; similar bores may be found in rivers elsewhere on the globe, such as Britain's Severn and China's Yangtze Kiang.

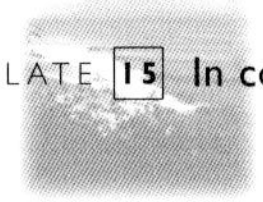
PLATE 15 In contrast with the self-contained pulse of a river bore, long-travelled Southern Ocean groundswells come into Indonesia's Desert Point in broad swathes, resulting in one of the world's finest "peeling" surfing waves.

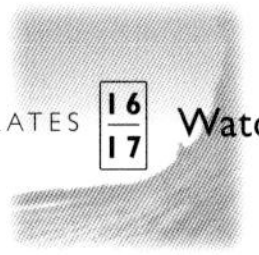
PLATES 16/17 Watched by a spurned surfer, a wave pitches into wintry morning air. The early land breeze — created when the air above the water grows warmer than that above the coastal land and rises above it as a result — can be seen breaking the lip line of the plunging wave into a hundred flinty fragments.

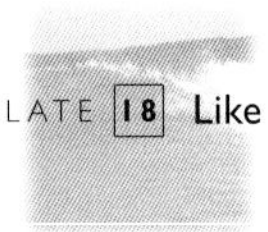
PLATE 18 Like a Japanese Bonsai plant, a miniature wave funnels its way down an inch-deep sandbar. The same dynamics govern waves of this size and monstrous storm-driven boomers; the scale may change, but the behaviour does not.

Experienced surfers become expert judges of relative breaker height, but 'professional' modesty shades their absolute estimates by about a factor of two... More precise estimates are easily made by anyone from shore by planting a pole in the sand, and marking the point on it that coincides with your line of sight across breaker crest to horizon. The elevation of this point above mean sea level can then be determined between groups of waves, when the sea is temporarily calm. Add 25 per cent for trough depression to create the wave height. I have done this many times in the presence of witnesses – but surfers are hard to convince.

William G. Van DORN

OCEANOGRAPHY AND SEAMANSHIP

1974

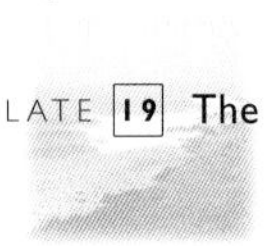
PLATE 19 The North Atlantic pours energy on to the desert coast of Morocco in north-west Africa. Temperate and tropical west coasts like this, the western USA, Chile, and West Australia take the brunt of the planet's swell power, as the major atmospheric weather systems march steadily from west to east around the globe.

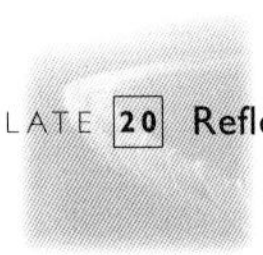
PLATE 20 Reflecting the isobar lines of its origin, a big cyclone-spawned swell peels down a remote central Queensland point. Cyclone swells, though intense, rarely travel far from the storm in question; the winds may be strong, but they rarely blow over the same stretch of water for long or across great distances. But within a thousand-kilometre range, the swell can still have quite an impact.

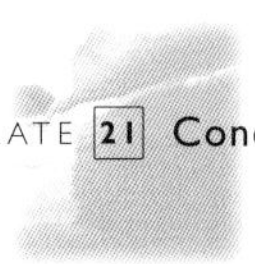
PLATE 21 Concaved to a fault, a thick-backed swell bends powerfully into a shorebreak in Hawaii. That raw energy, breaking safely a dozen metres offshore, is unlikely to damage anything but itself; a surfer or swimmer nearby would have to play it safe to avoid being caught up in its savage ending.

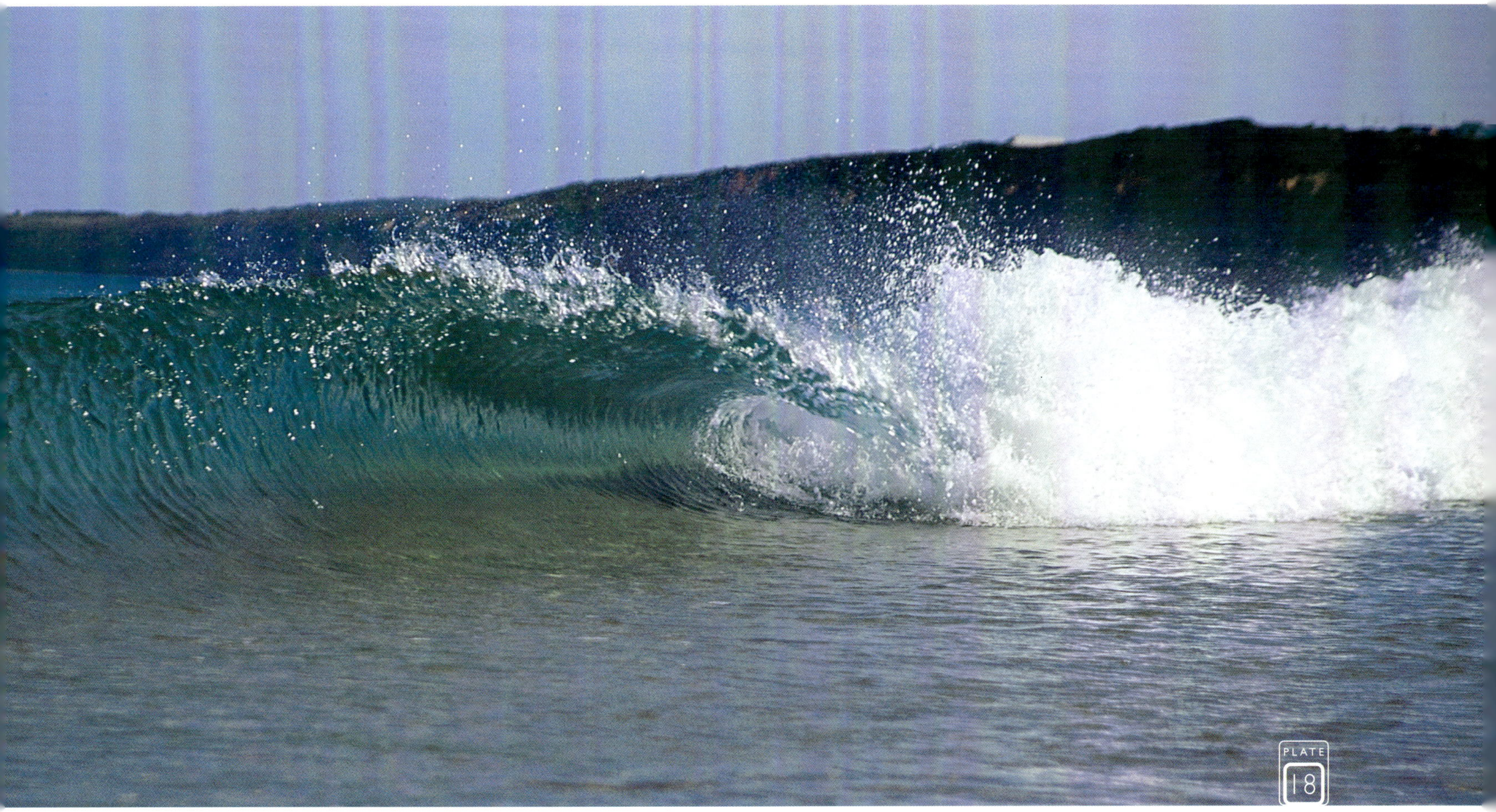
PLATE
18

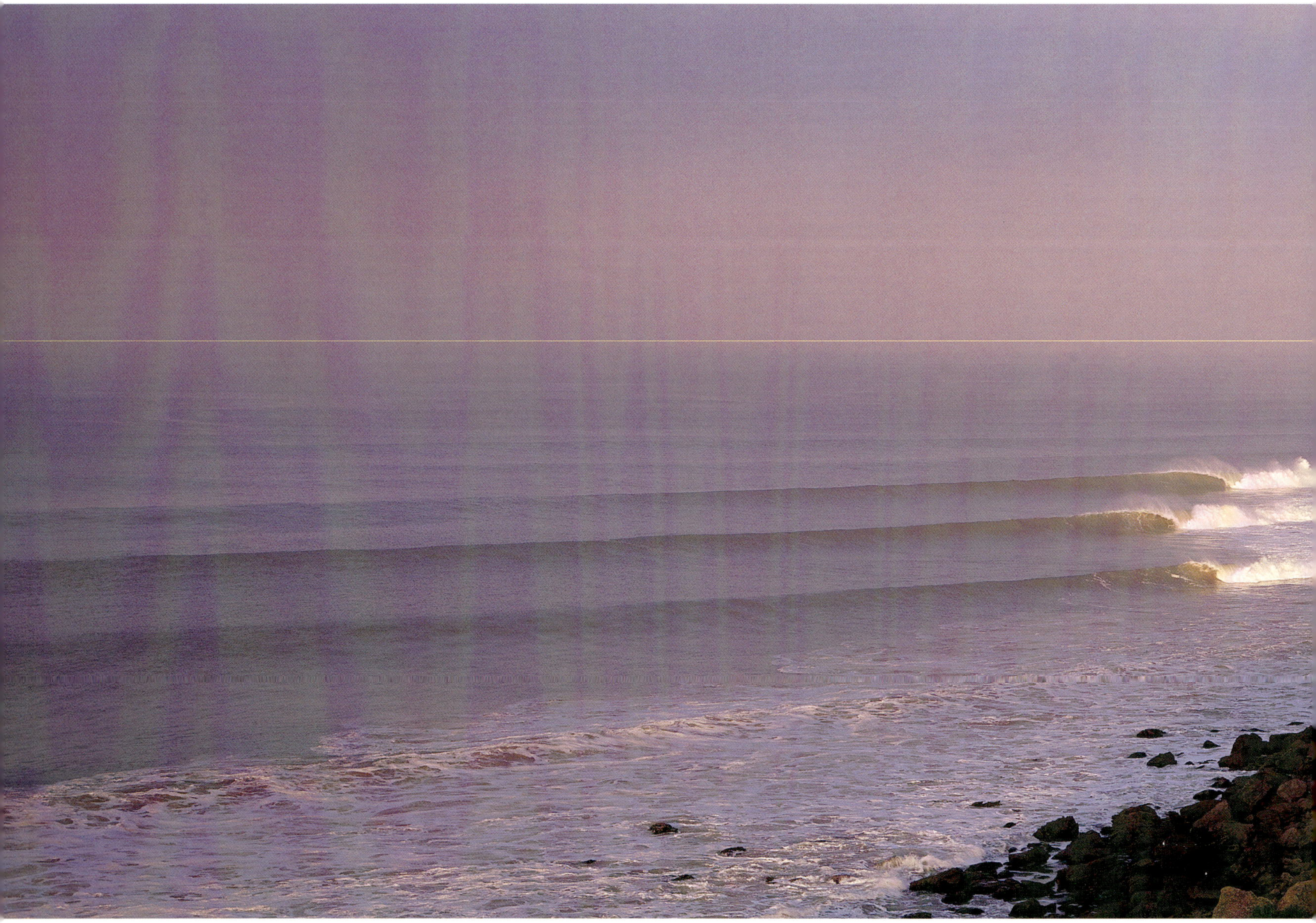

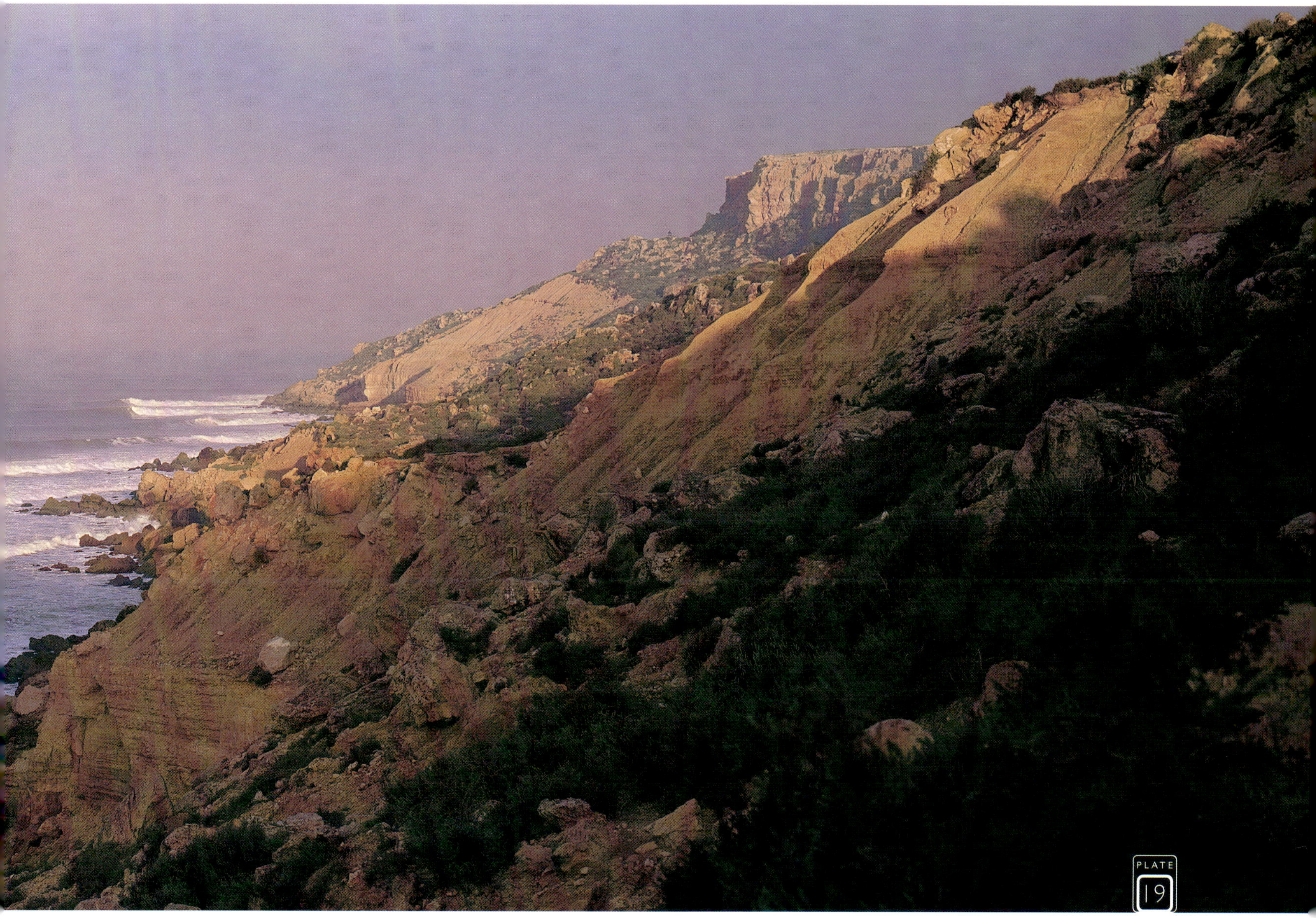

PLATE 19

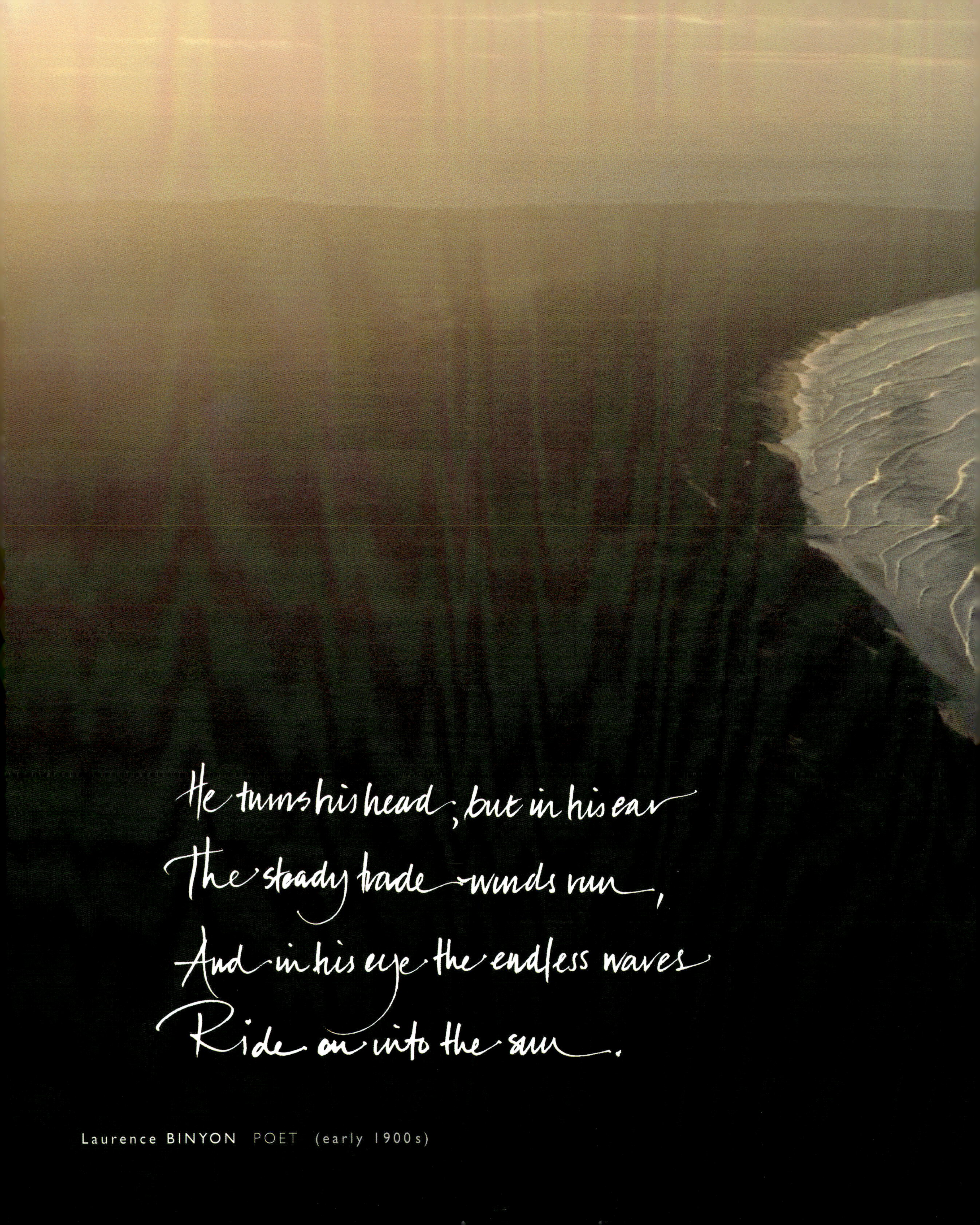

He turns his head; but in his ear
The steady trade winds run,
And in his eye the endless waves
Ride on into the sun.

Laurence BINYON POET (early 1900s)

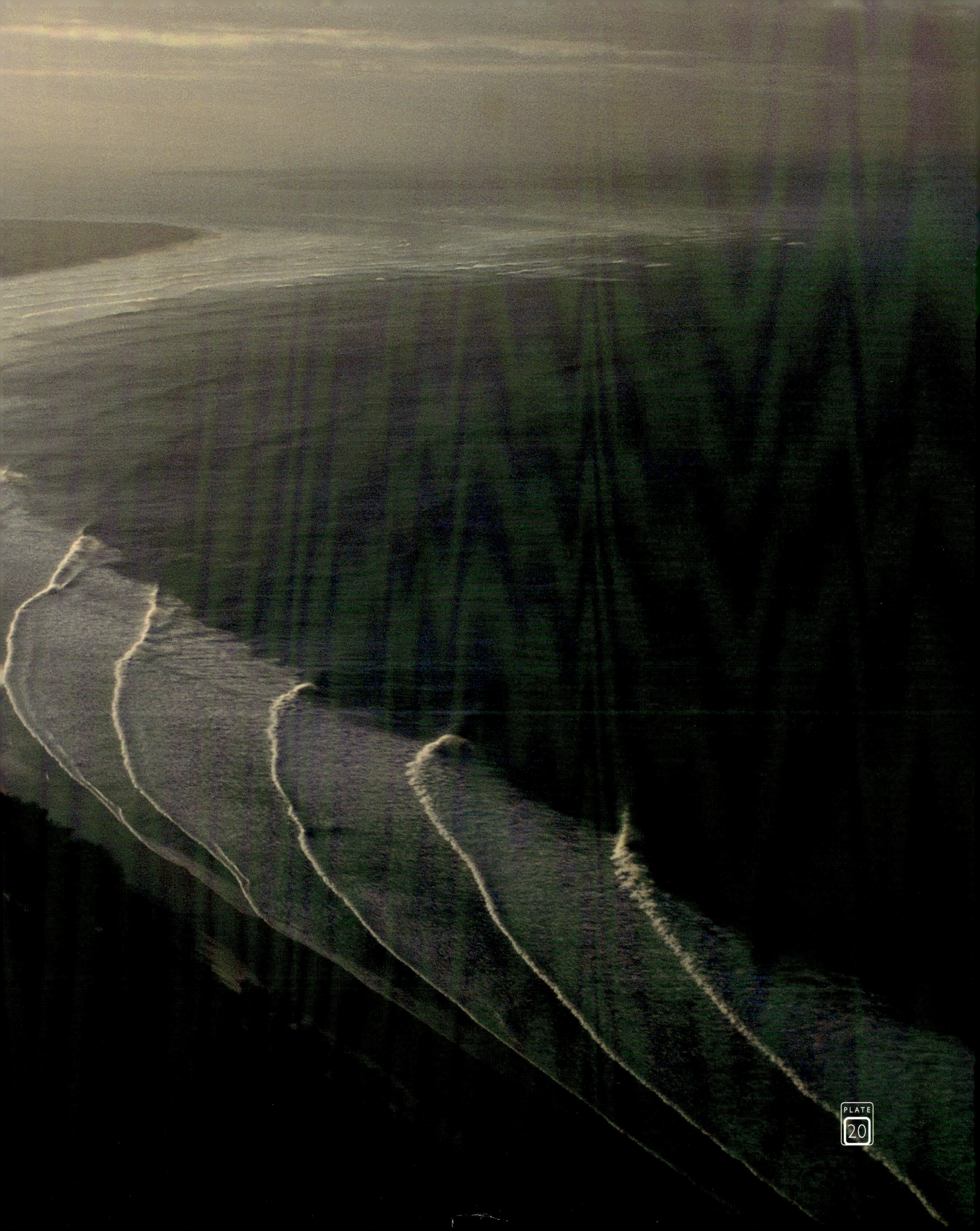

PLATE
20

Plunging breakers are the most impressive. Their principal characteristic is very rapid release of energy from a wave moving at high velocity. There is a sudden deficiency in water ahead of the wave which causes high-velocity currents in the trough as the water rushes seaward to fill the cavity beneath the oncoming crest. When there is not enough water to complete the wave form, the water in the crest, attempting to complete its orbit, is hurled ahead of its steep forward side and lands in the trough. This curling mass of falling water will often entrap air and then, as the upper part of the wave collapses, the air is compressed. When the compressed air finally bursts through the watery cap, a geyser of water is hurled into the air—sometimes over fifty feet.

Willard **BASCOM** WAVES AND BEACHES 1958

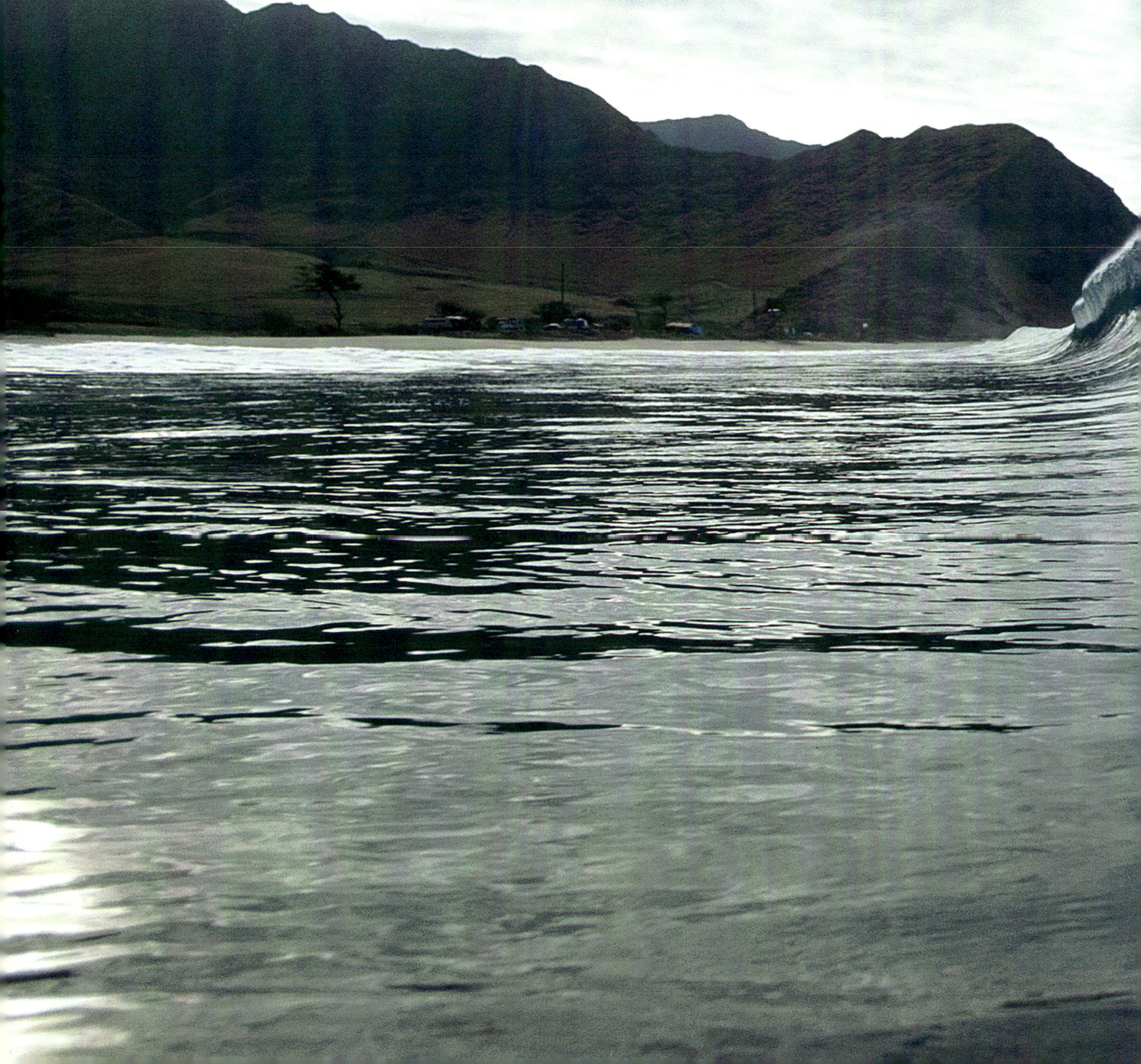

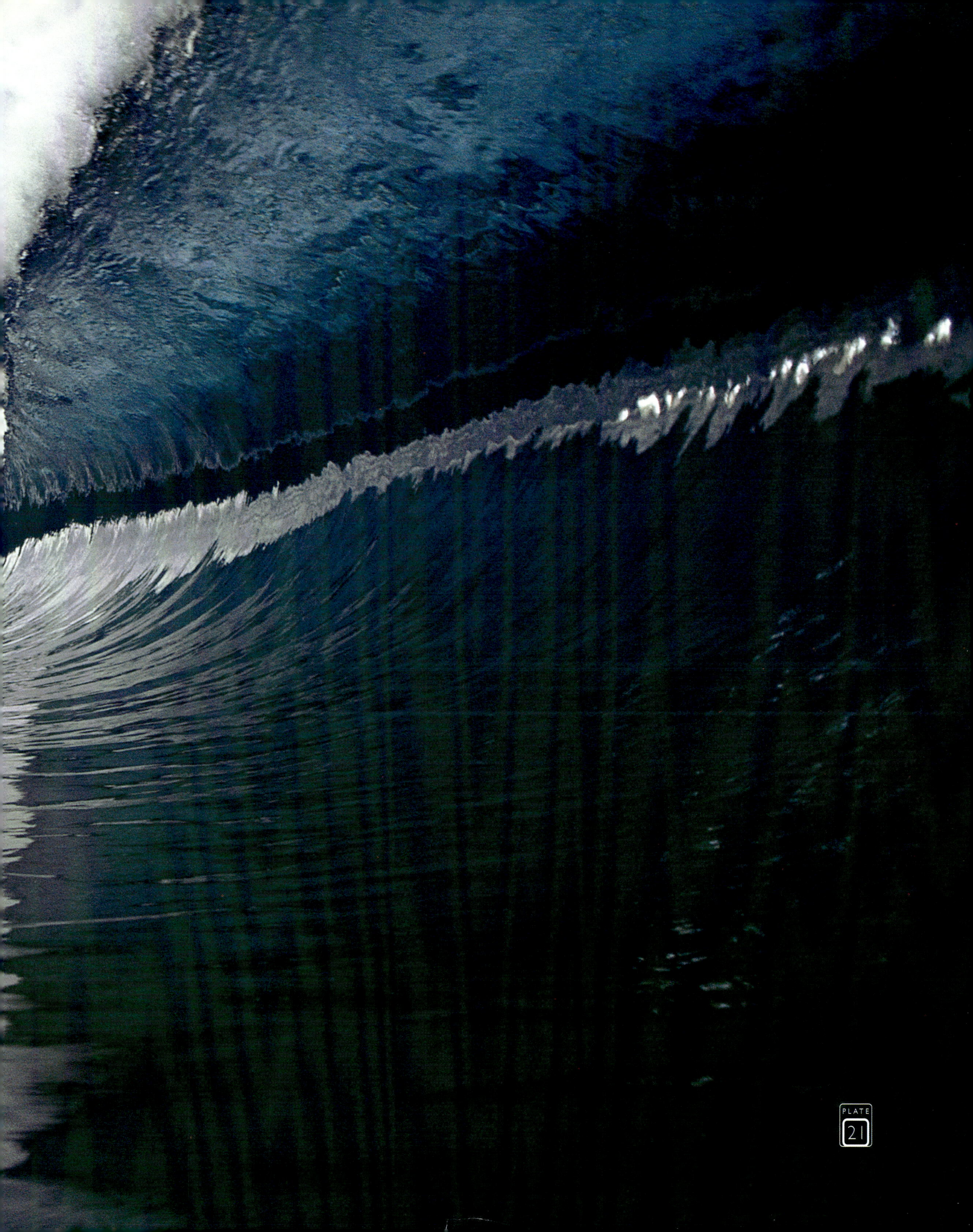

PLATE 21

That instant of rushing towards the beach pushed along by the bubbling whitewater strikes a primordial chord deep inside of us that goes "This is unreal!" And so begins, for many of us, the lifelong search for the Holy Grail — the perfect wave. It begins with freezing pre-dawn expeditions to the local beach, scouring the ocean for any morsel it may serve up. Do you remember going out in those two-foot closeouts for hours, looking for that one wave that would hold up? Or the days sitting, watching, unable to deal with the power and intensity the ocean was capable of. Watching in awe at a better surfer's seemingly effortless mastery of them. Unable to look at the ocean passively, any glimpse you get is an anxious search for the possibility of getting a good wave. "What's the wind doing? When's mid-tide? Is that sandbar still holding up? I wonder if it's hitting the reef at the right angle."

It becomes a game of time and space, learning to intersect with a pulse in the ocean that may have travelled thousands of kilometres, at the precise moment, in that exact space where a green face will hold up and offer opportunity to revel in that weightless feeling of skimming across the water.

Be a couple of seconds out or two metres from that precise point and it's no dice! Like an intergalactic spaceship searching for a hole in the time continuum of space, we move around the watery gameboard, attempting to line up that precise time/space point where a surfable wave will be created. And we are offered a smorgasbord of different varieties — the long gentle peeling walls, the fun glassy peaks, intense spitting barrels, windblown chopped-up summer mush... It becomes an obsession that non-surfers cannot comprehend. And if by some chance you are reading this and have never caught a wave, be warned — catch a wave and your life may never be the same again.

Jim BANKS AUSTRALIA'S SURFING LIFE MAGAZINE 1995

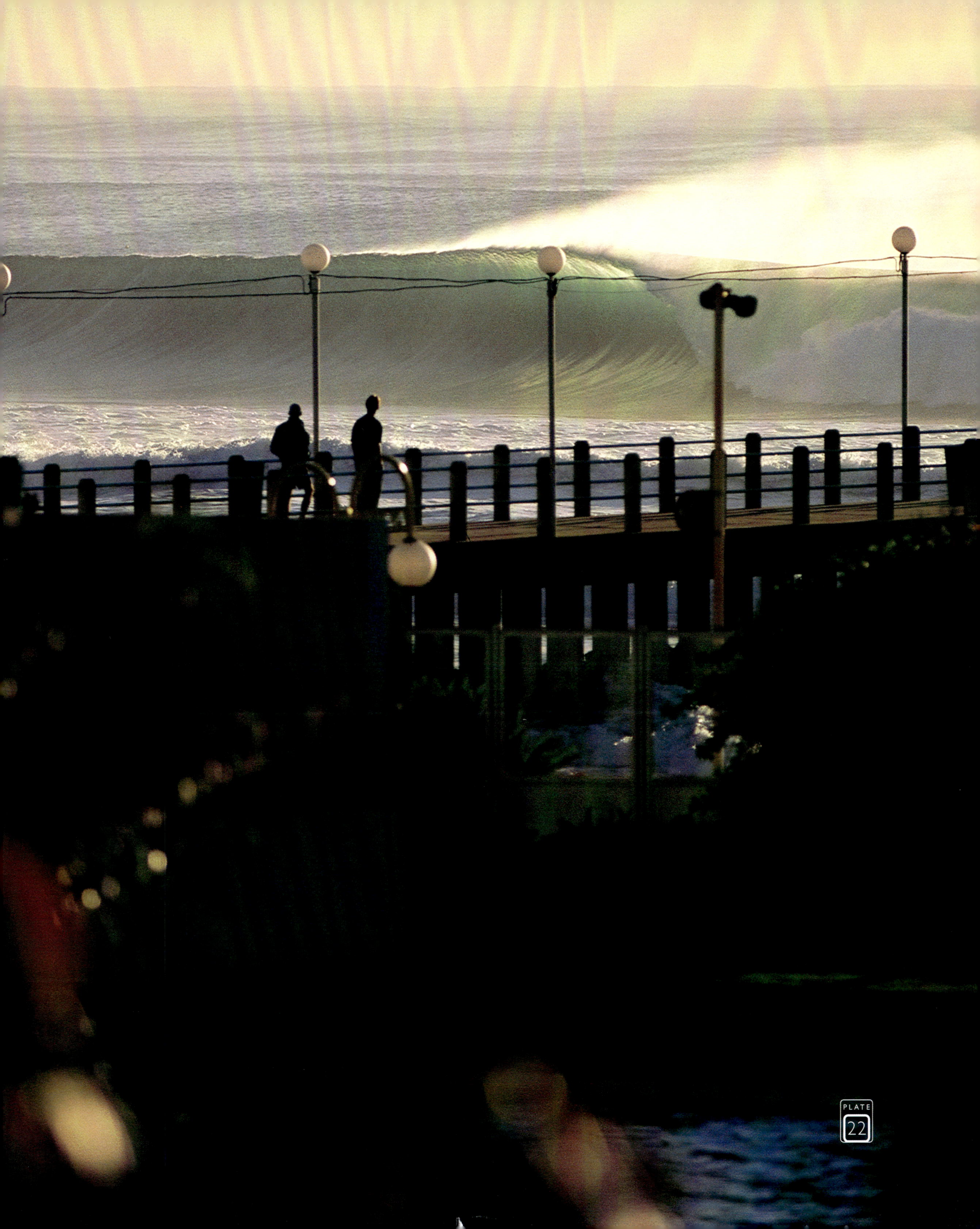

PLATE 22

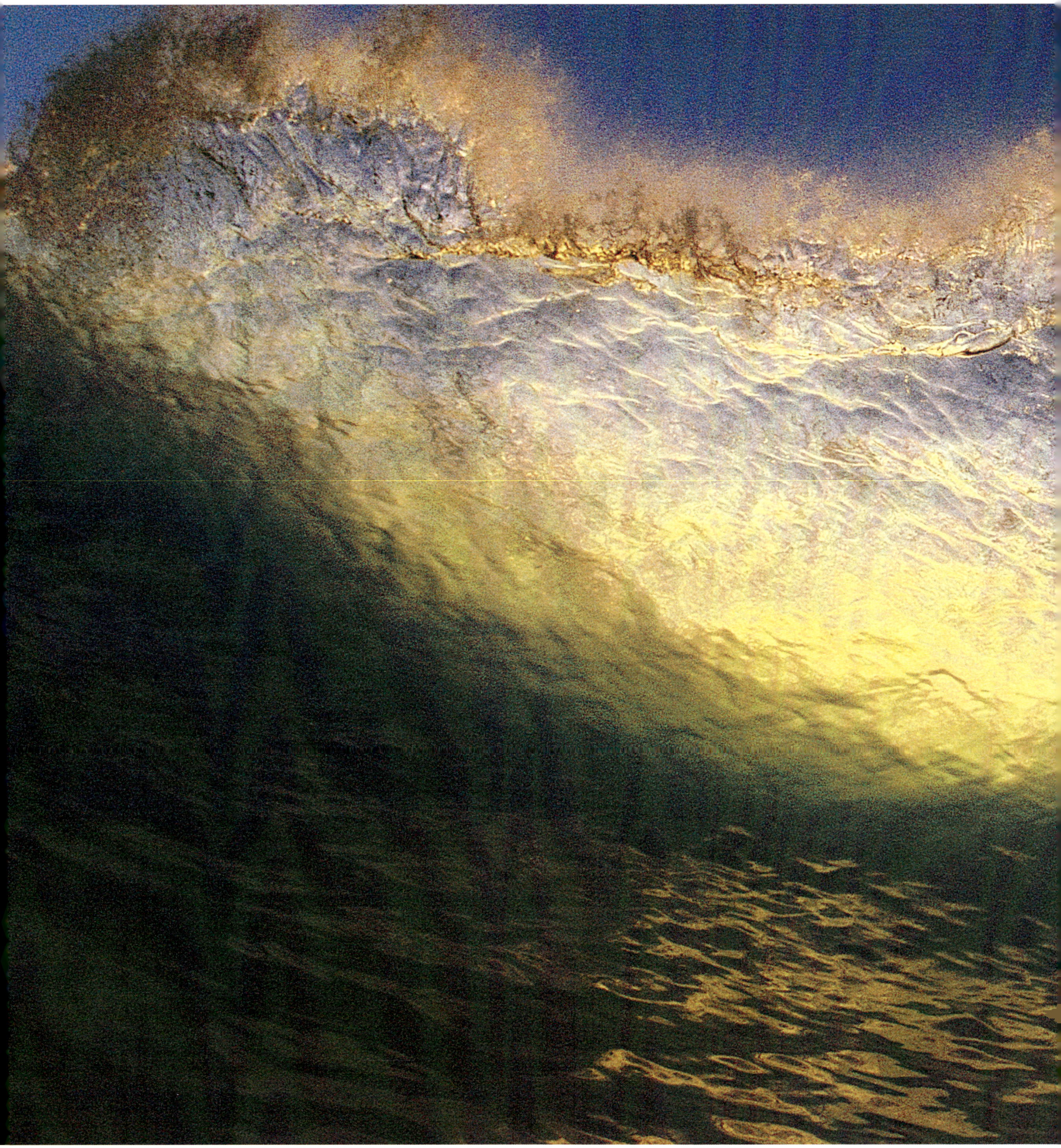

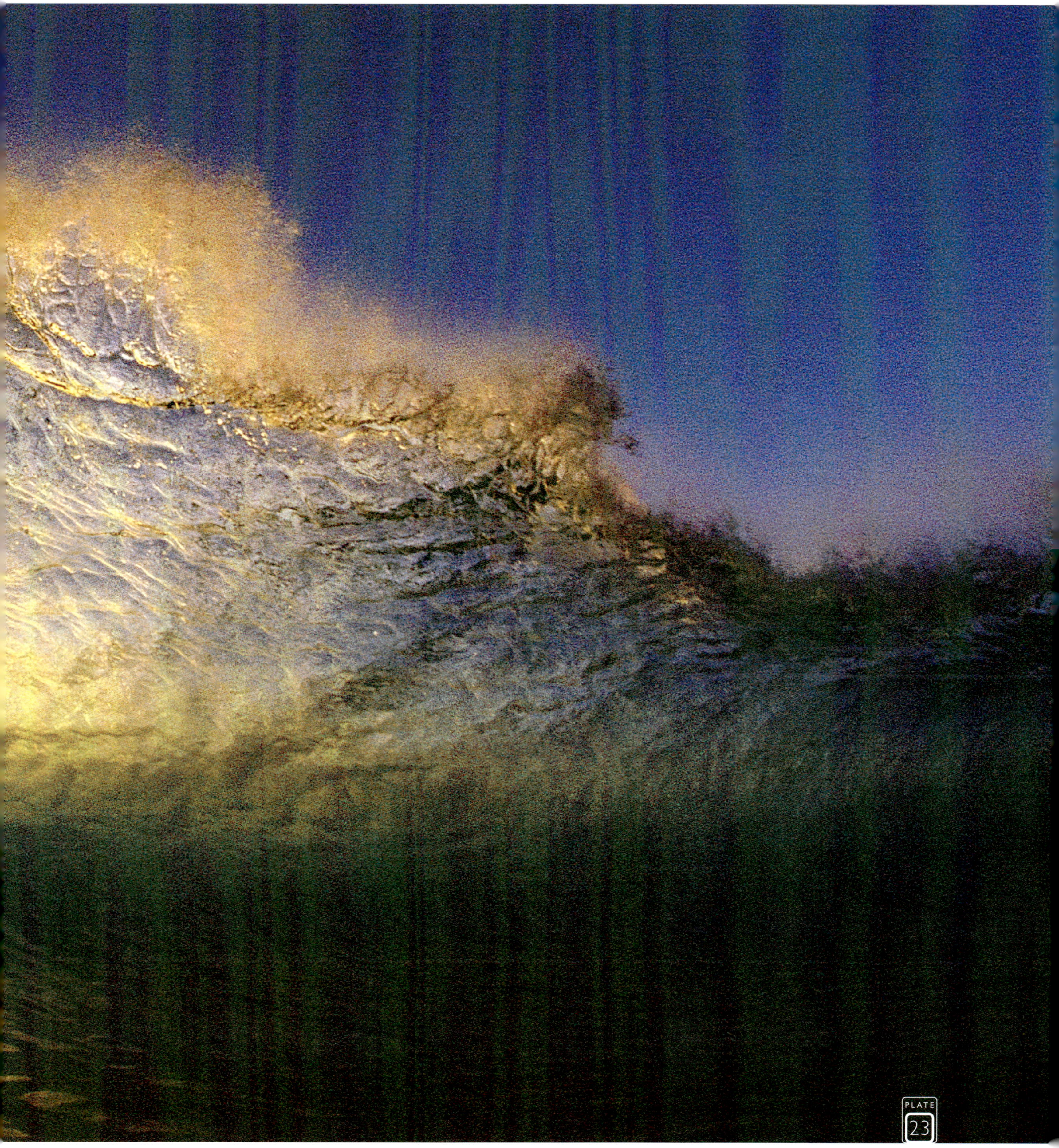
PLATE
23

Who hath desired the Sea? — the sight of salt water unbounded —

– The heave and the halt and the hurl and the crash of the combered wind-hounded?

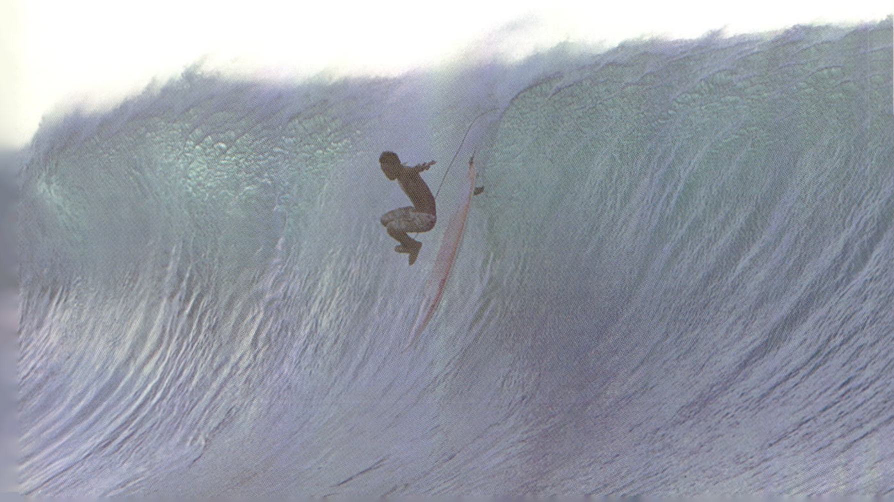

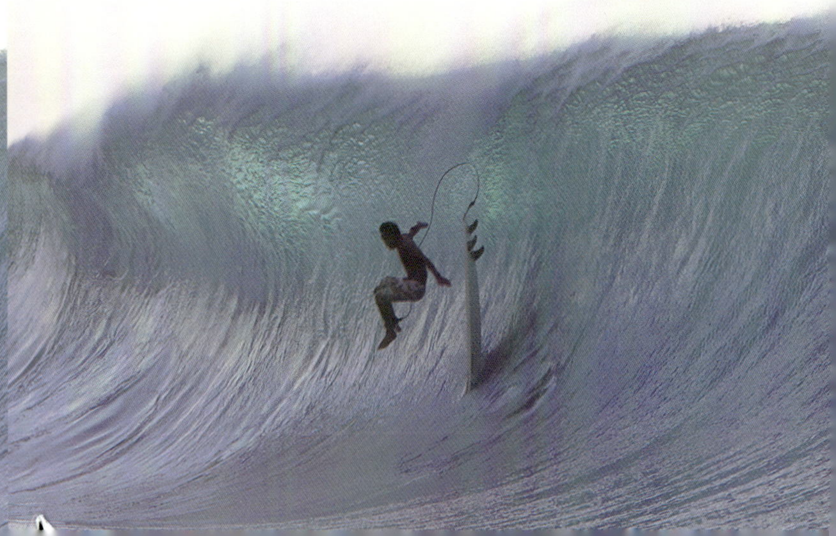

The sleek-barrelled swell before storm, grey, foamless, enormous, and growing ...

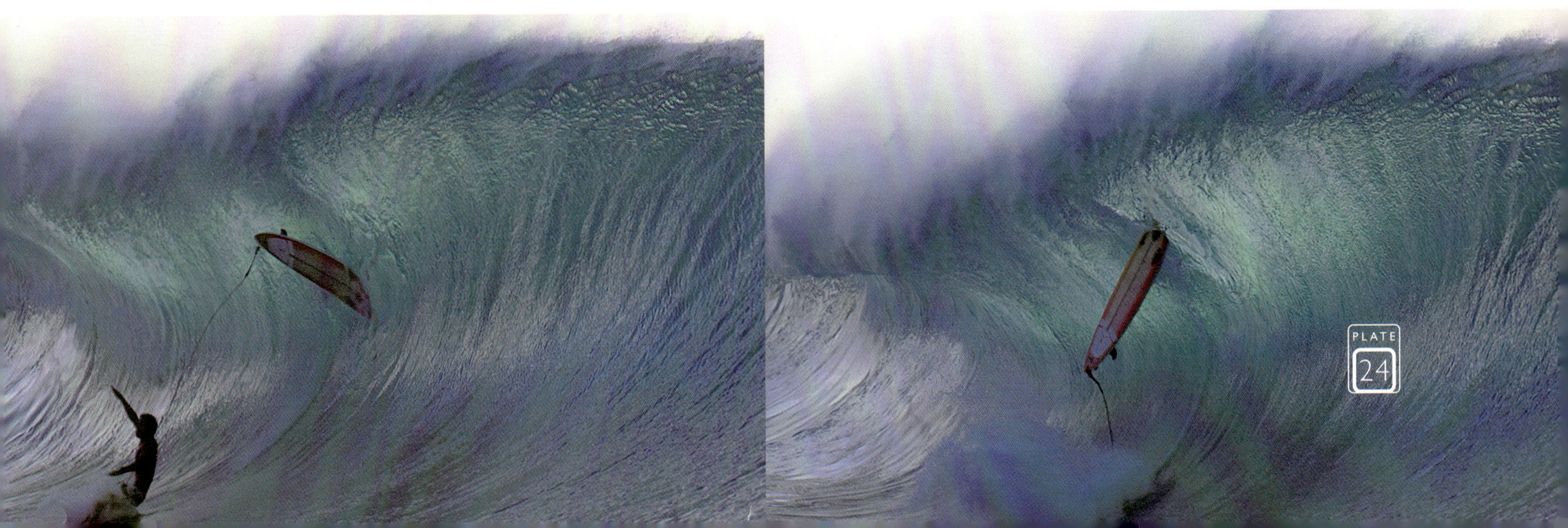

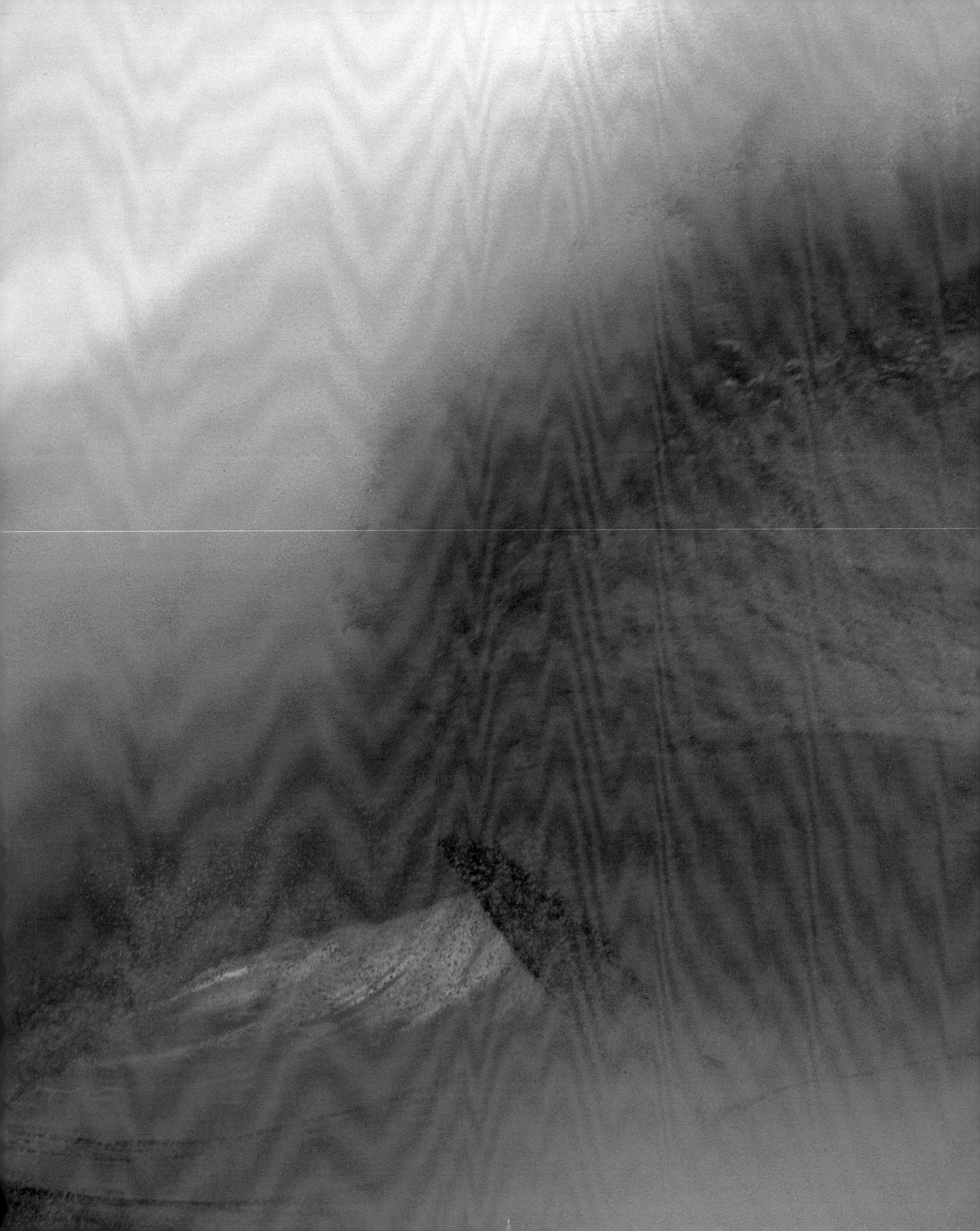

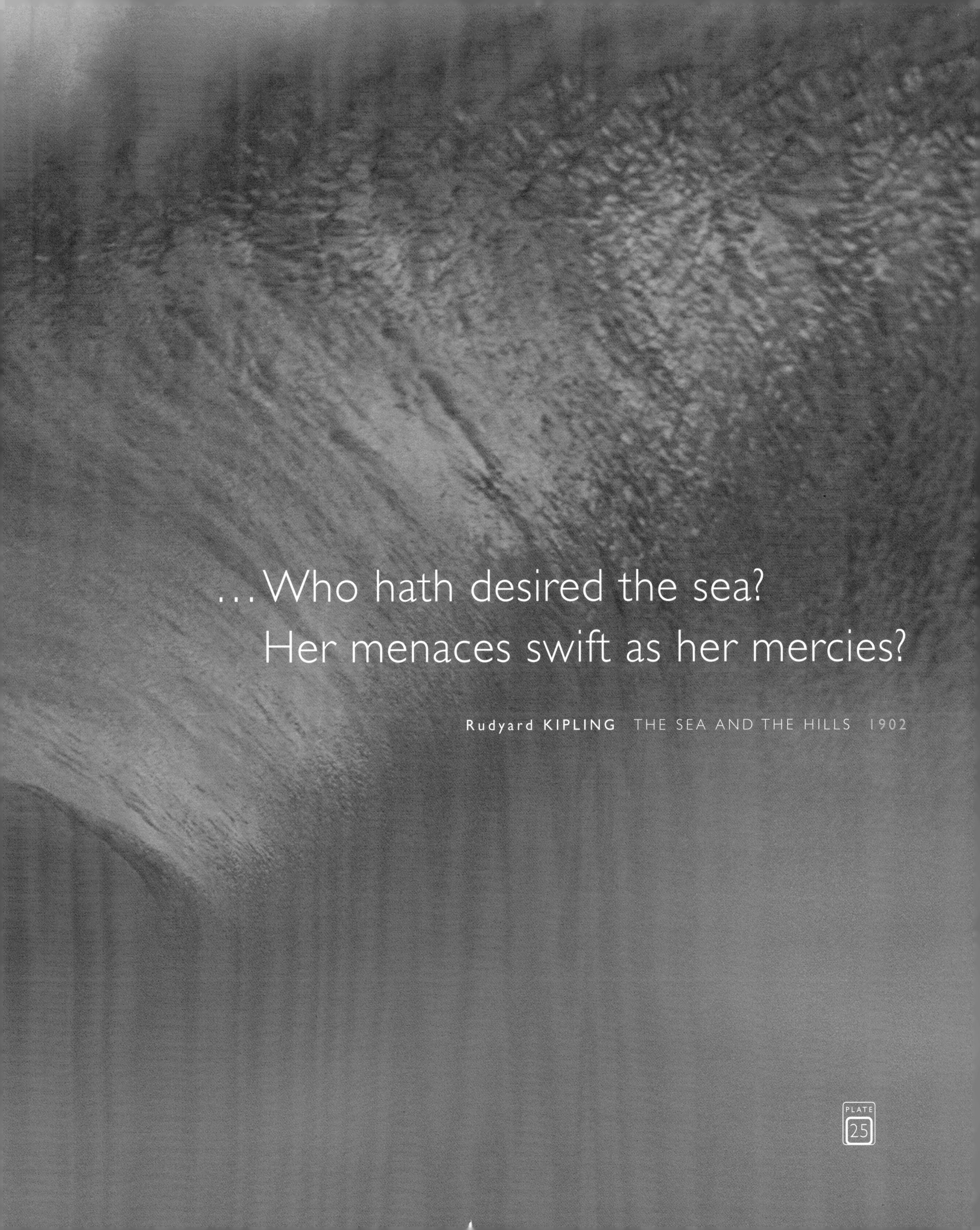

…Who hath desired the sea?
Her menaces swift as her mercies?

Rudyard **KIPLING** THE SEA AND THE HILLS 1902

PLATE 25

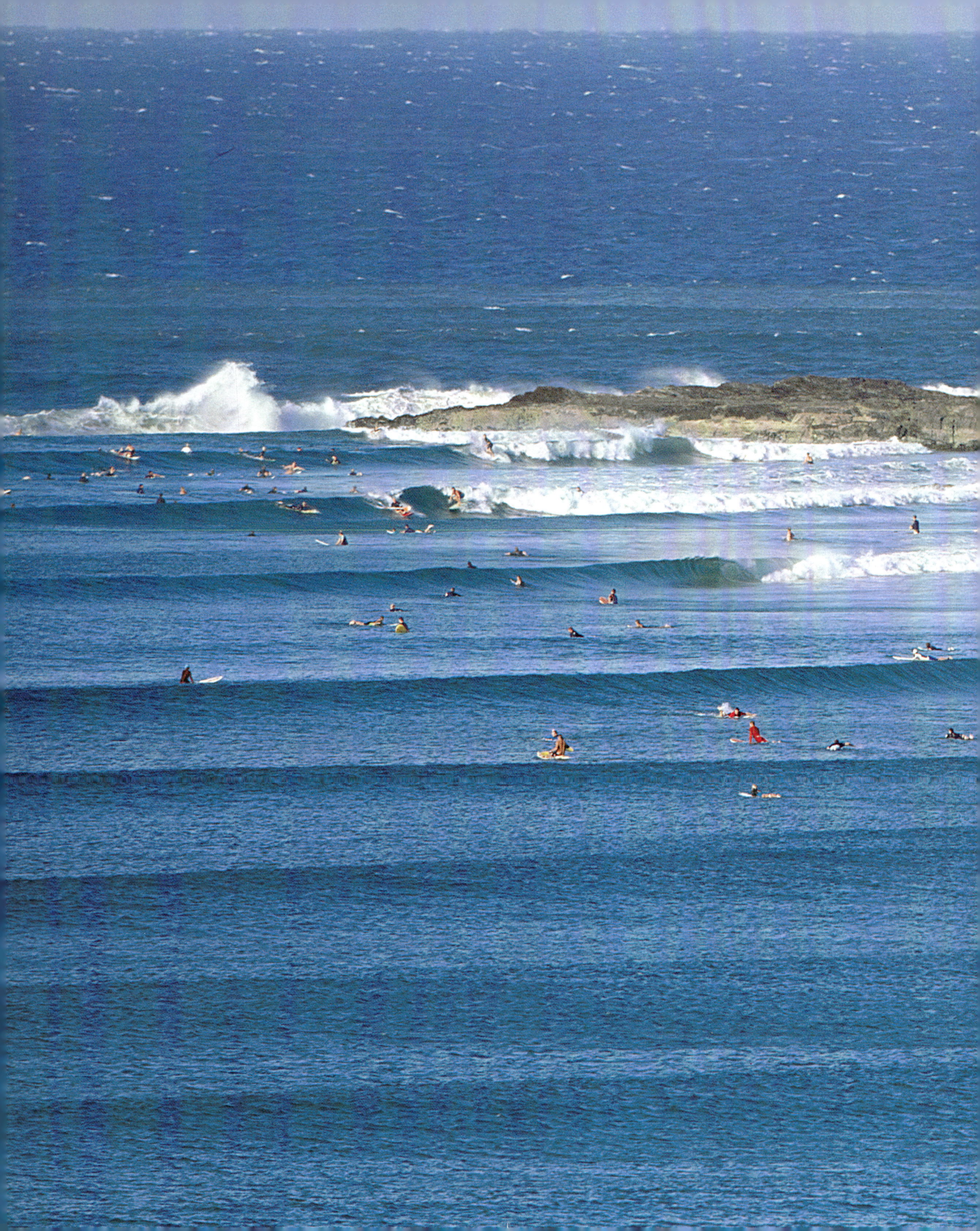

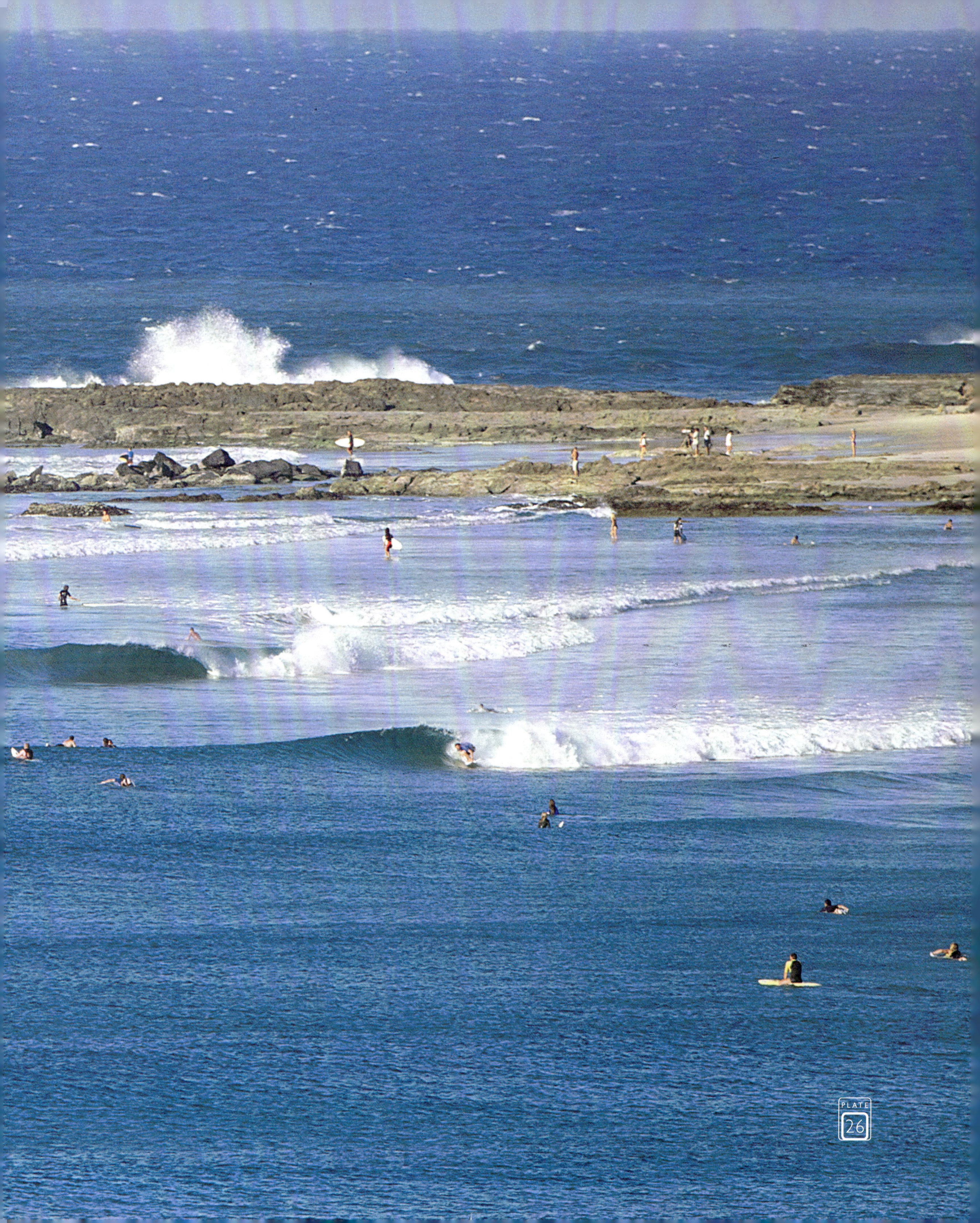

PLATE
26

For just as the dolphin glides
through the water on swift fins,
now rising above the surface and
now sinking to the depths, and
piles up waves and sends them
off in circles, just so will each
person born under the sign of
the Dolphin fly through the waves
raising one arm and then the
other in slow arcs.

MANILIUS ROMAN POET *first century A.D.*

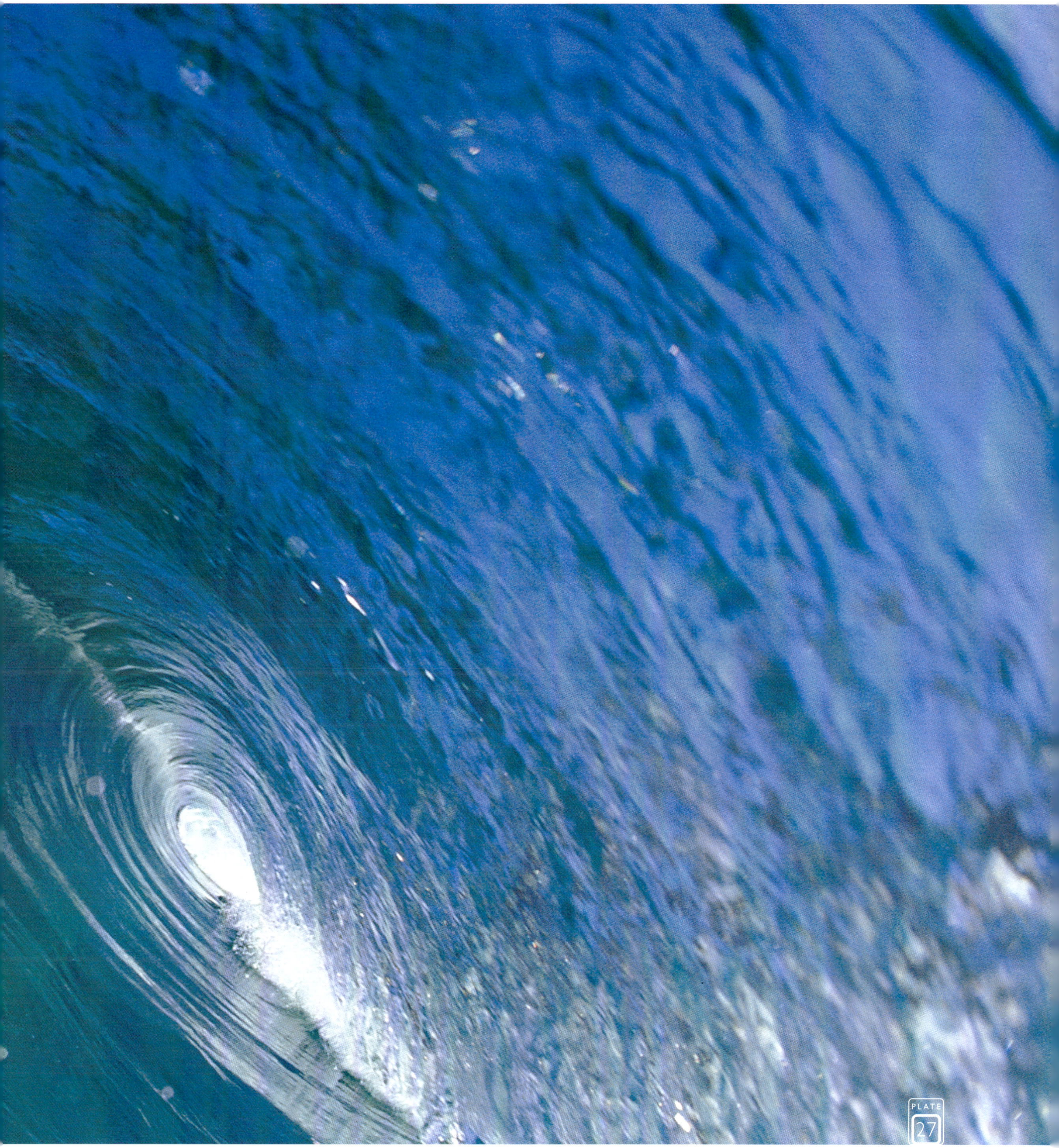

PLATE
27

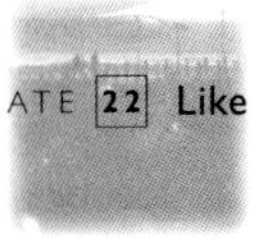

PLATE **22** Like a freight train blowing smoke, a big wave roars down past New Pier, Bay of Plenty, Durban, South Africa. Piers can create excellent waves for surfers by creating a focal point for wave action on an otherwise straight-lined beach.

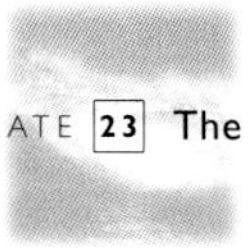

PLATE **23** The lacy face of a tropical swell captures the morning sun as it pitches and falls. Tropic water colour — that rich transparent blue-green — often belies the inner power of waves that break hard on coral and lava reefs.

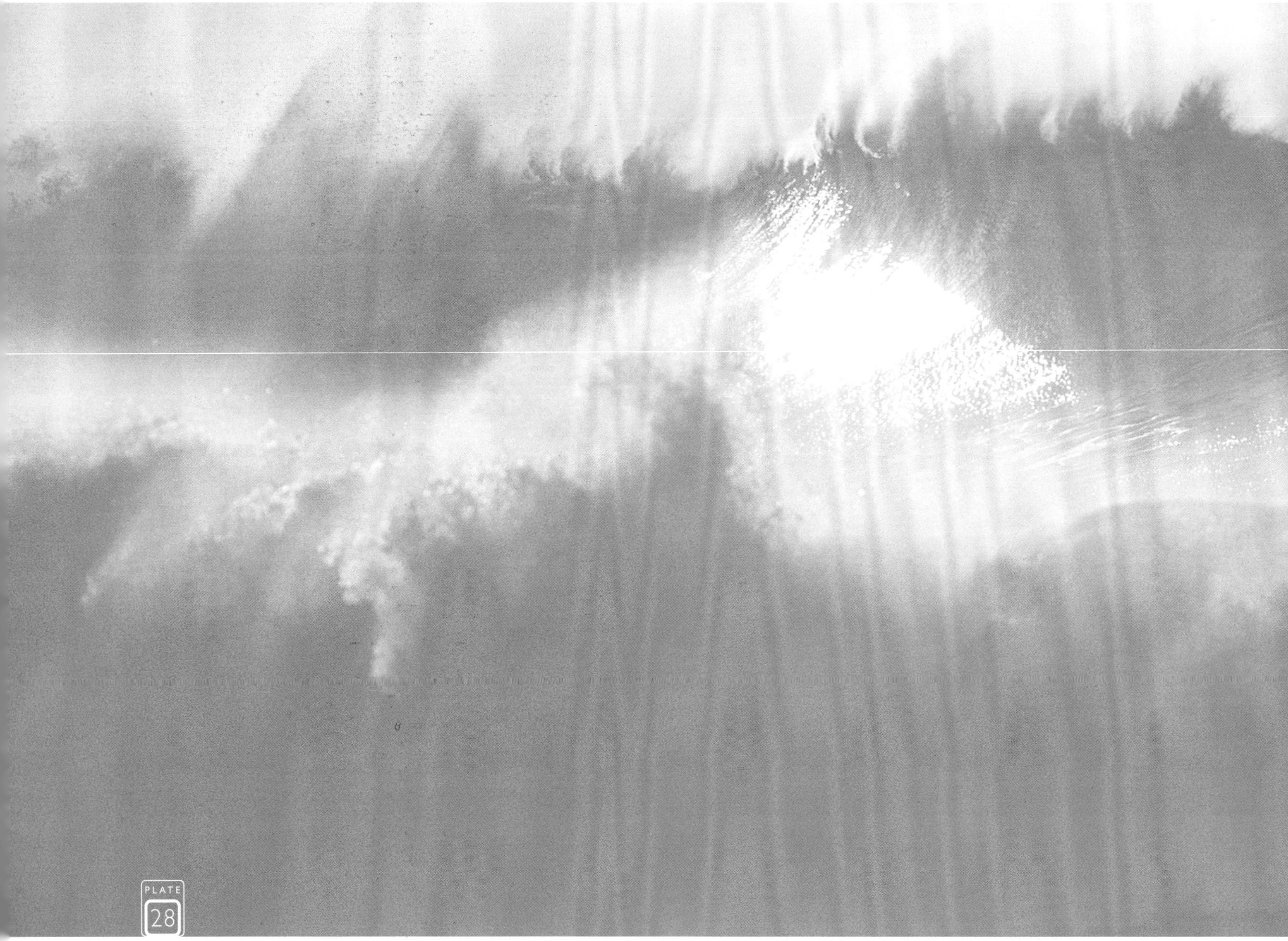

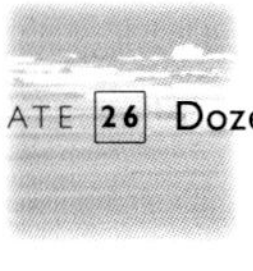

PLATE **26** Dozens of keen surfers flock to the small clean waves inside Snapper Rocks and Rainbow Bay on Queensland's Gold Coast. The popularity of surfing has grown rapidly worldwide in recent decades, as more and more people shift to coastal lifestyles in countries from Europe to Asia and beyond — not to mention in Australia, one of the homelands of coastal culture. It's to be hoped that as we grow in our enjoyment of the coasts and waves of the world, we don't lose sight of the need to protect them.

PLATES 24 25 Young Hawaiian surfer Tory Barron finds himself caught within the fearsome Banzai Pipeline cauldron. Note the vicious compression of energy against the suddenly rising bottom contour, actually driving the lip of this ferocious wave skyward before its inevitable explosive collapse. (Somehow Tory came through without a scratch.)

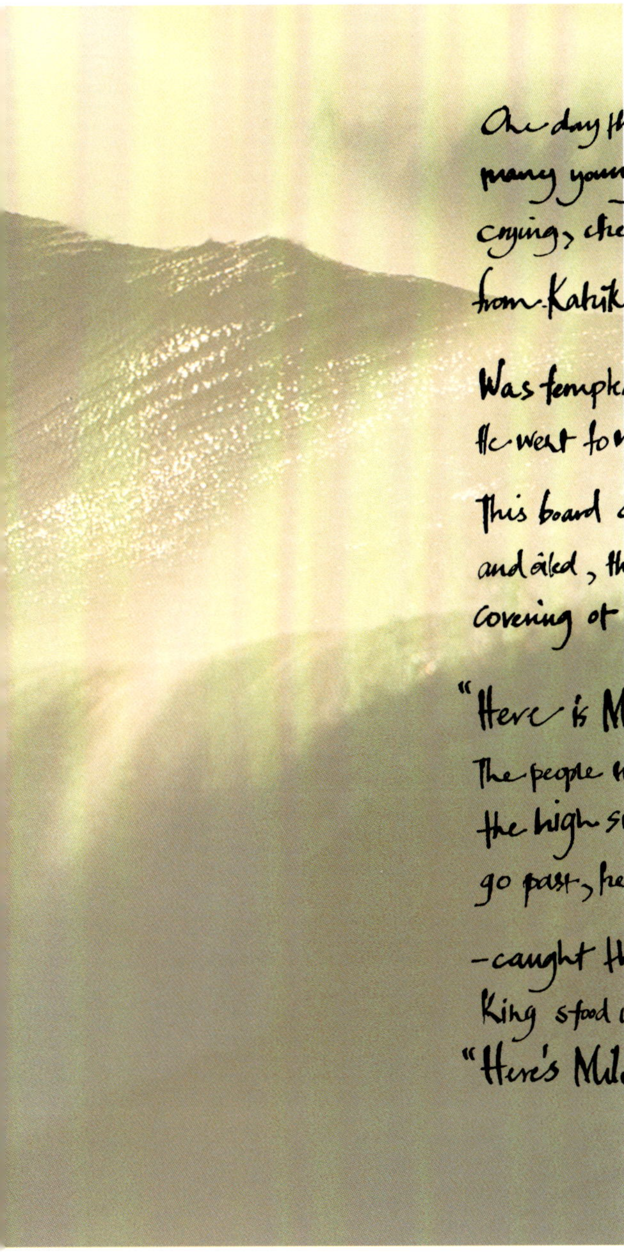

One day the surf at Waipio was high. Great combers rolled upon the beach and many young chiefs had their surfboards out, the common people also. The crowds were crying, cheering when the surfers rode. The riders said their chants for great waves from Kahiki and the gods replied with combers, long slow-rolling combers good to stand on.

Was tempted Milu then. He left his healing house, he lifted up the leaves; came out. He went to where his board was kept, his chiefly olo made of Wiliwili.

This board of his he took, his surfboard that was longer than two tall men — well stained and oiled, that shining board. Milu took down his board from the rafters, unwrapped its covering of tapa, ran to the beach, swam out.

"Here is Milu! Here is Milu the King of his olo! Let us watch him."
The people watched their King seek out the place. He swam beyond the low surf, through the high surf, found the place; he waited. Then Milu let the first great wave go past, he let the second go. He saw his chance, he started on the third, a beauty!

– caught the crest. His chiefly board held up its head and handsome Milu stood, the King stood up and rode that rolling crest. The people cheered, they shouted, cried;
"Here's Milu! He's recovered! He is riding the surf!"

From LONOPUHA (or the origin of healing)

POLYNESIAN MYTHS AND LEGENDS

PLATE 27 Deep within a small tubular wave, airy foam gathers. Waves like this sometimes act almost as pressure cookers, squirting volumes of air and foamy spray high into the air or out through the tube as they collapse — a phenomenon known to surfers as "spit".

PLATE 28 A strong tradewind begins the process of cleaning up storm surf off Oahu. Waves like these are generated so close to shore that the process of swell decay has yet to fully smooth them into long even lines. The result: surf that can be unpredictable — and dangerous.

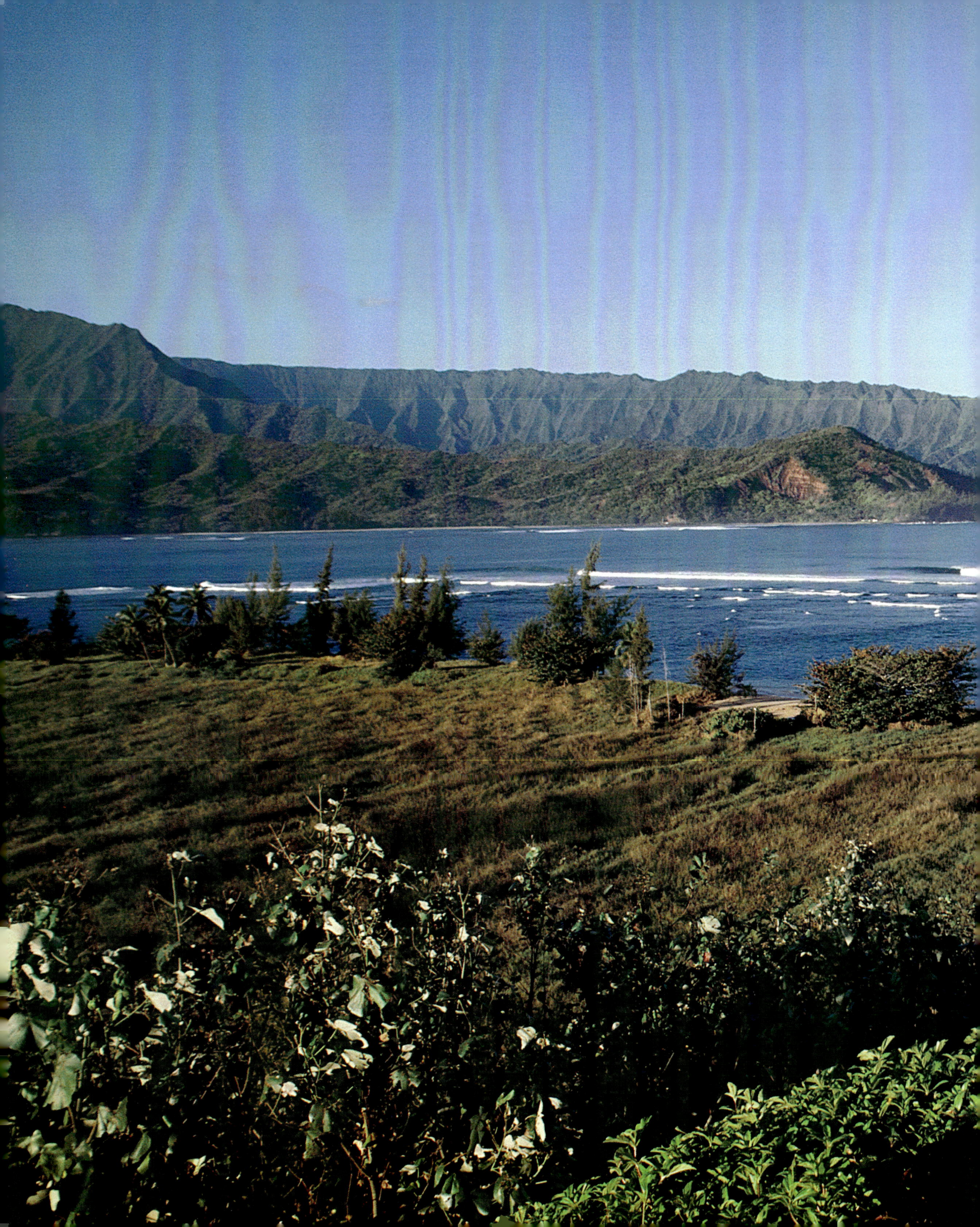

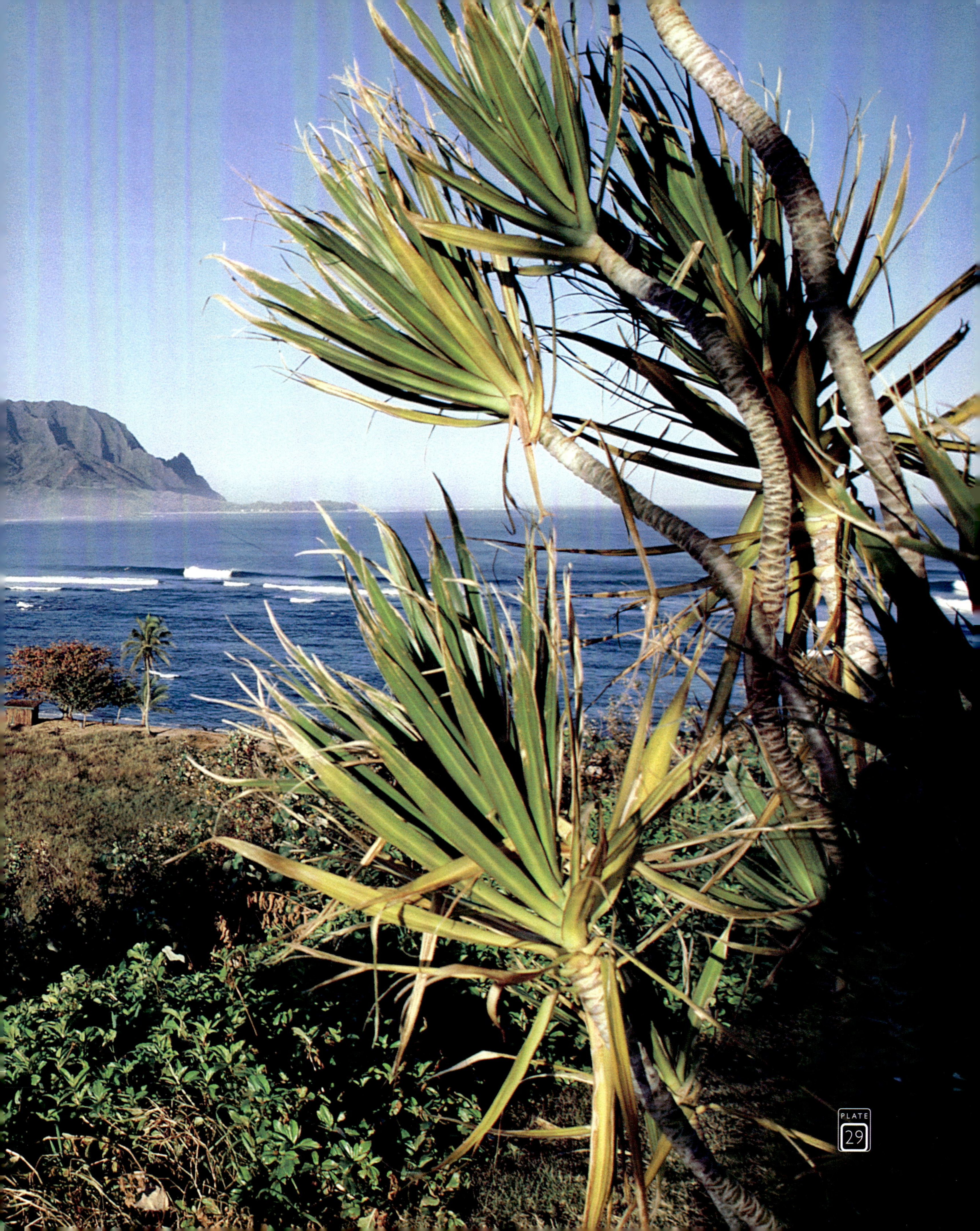

PLATE
29

Aug 10. I was looking at high waves. The breakers always are parallel to the coast and shape themselves to it except where the curve is sharp however the wind blows. They are rolled out by the shallowing shore just as a piece of putty between the palms whatever its shape runs into a long roll…

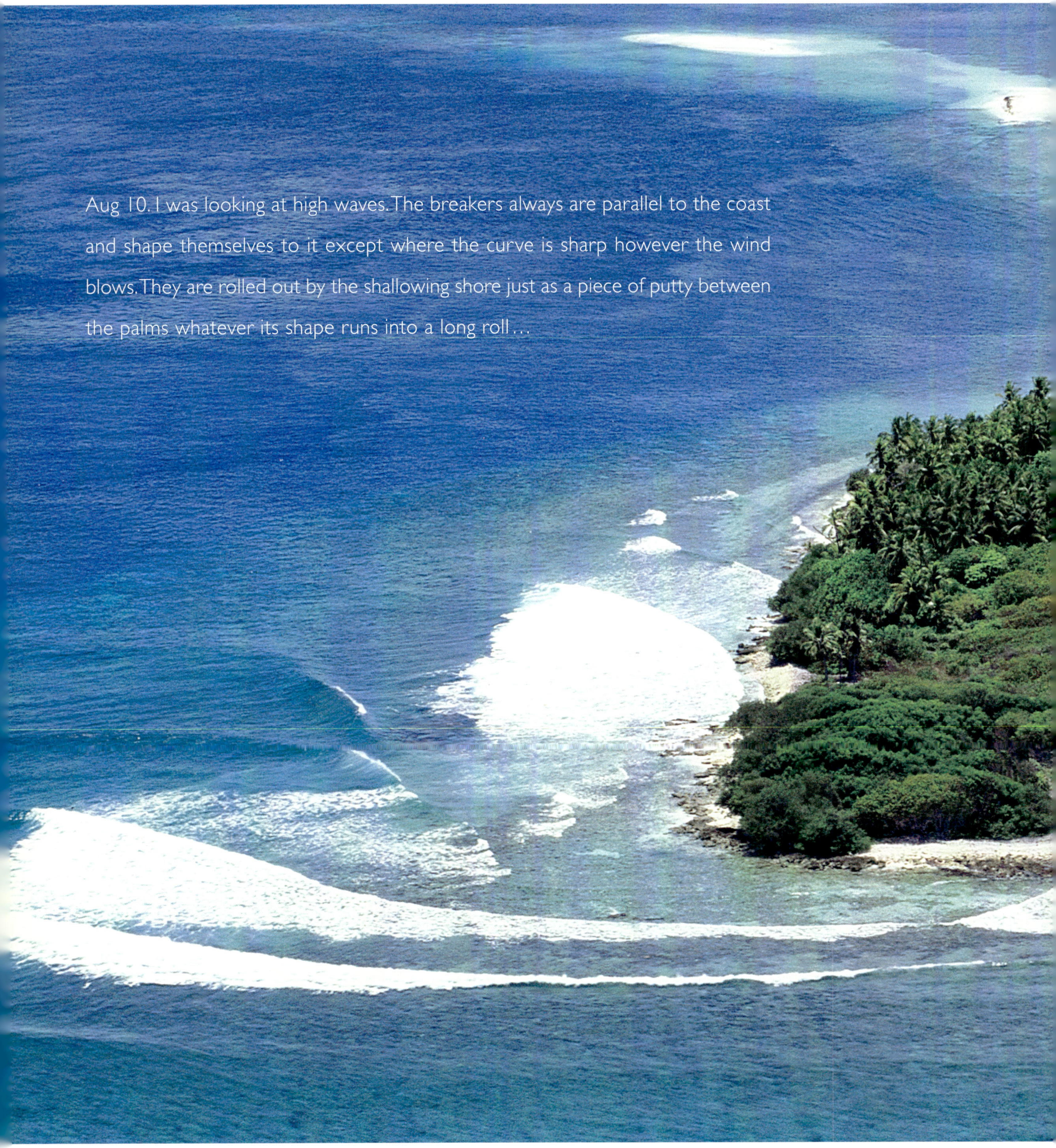

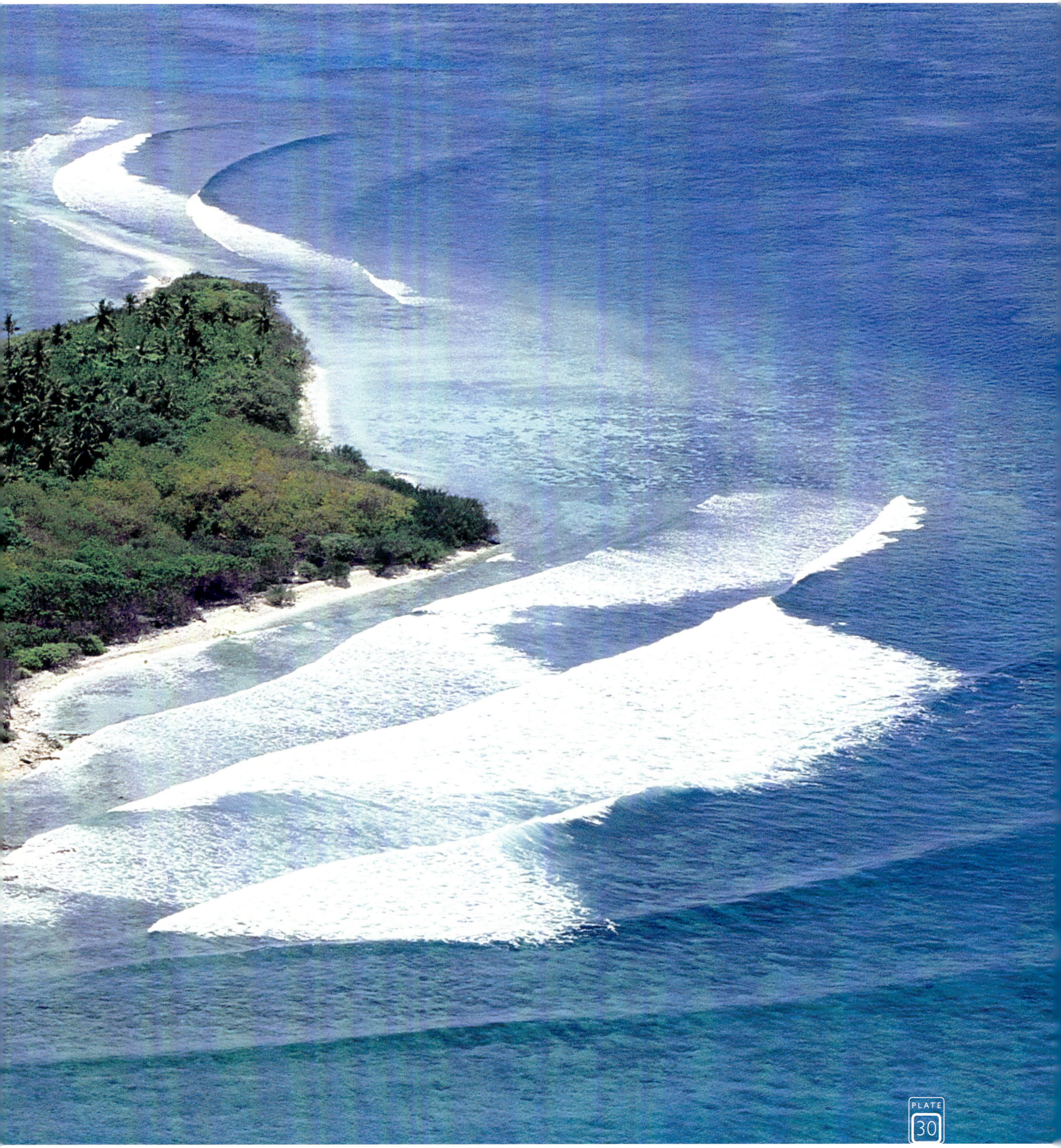

PLATE 30

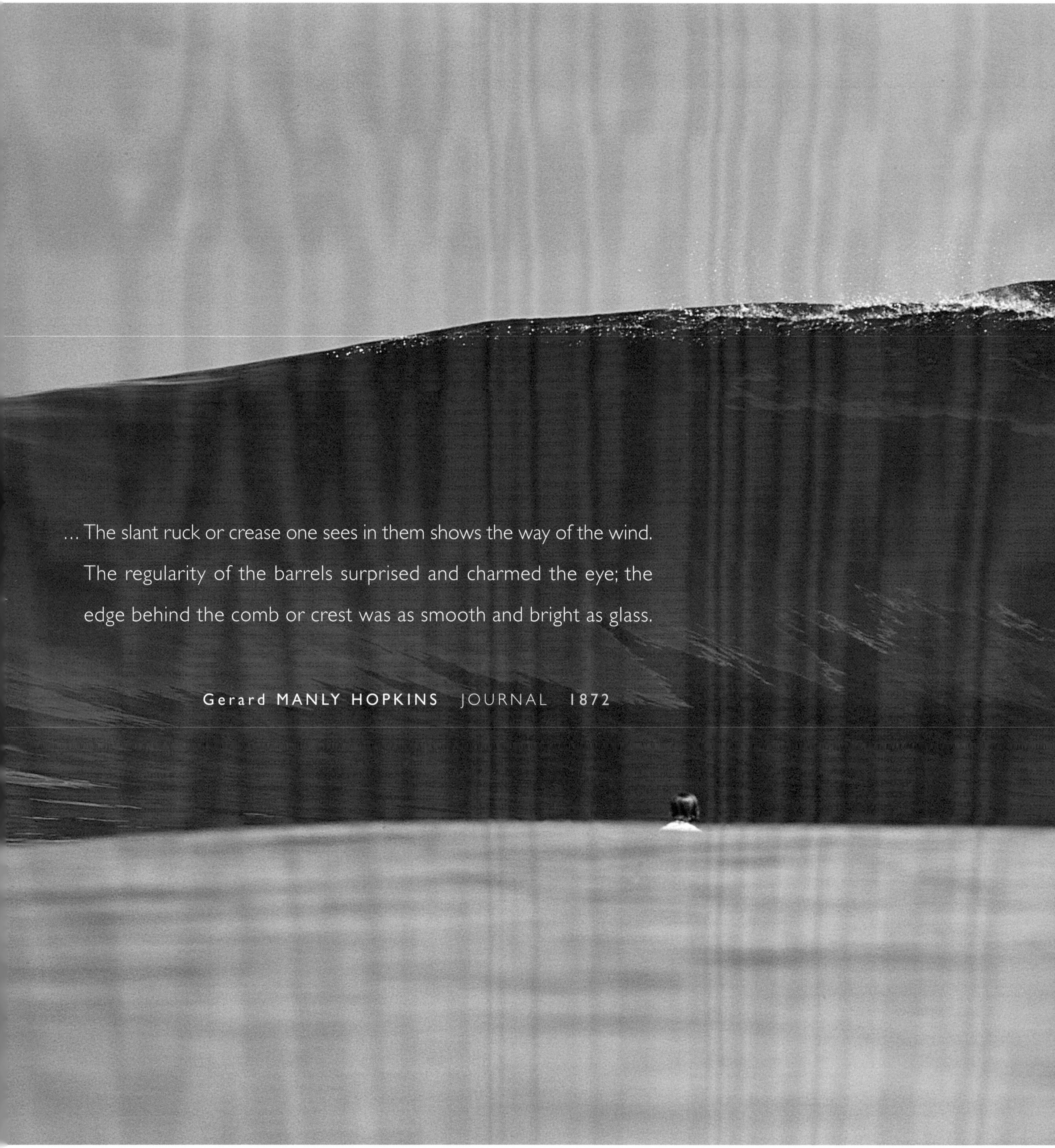

… The slant ruck or crease one sees in them shows the way of the wind. The regularity of the barrels surprised and charmed the eye; the edge behind the comb or crest was as smooth and bright as glass.

Gerard **MANLY HOPKINS** JOURNAL 1872

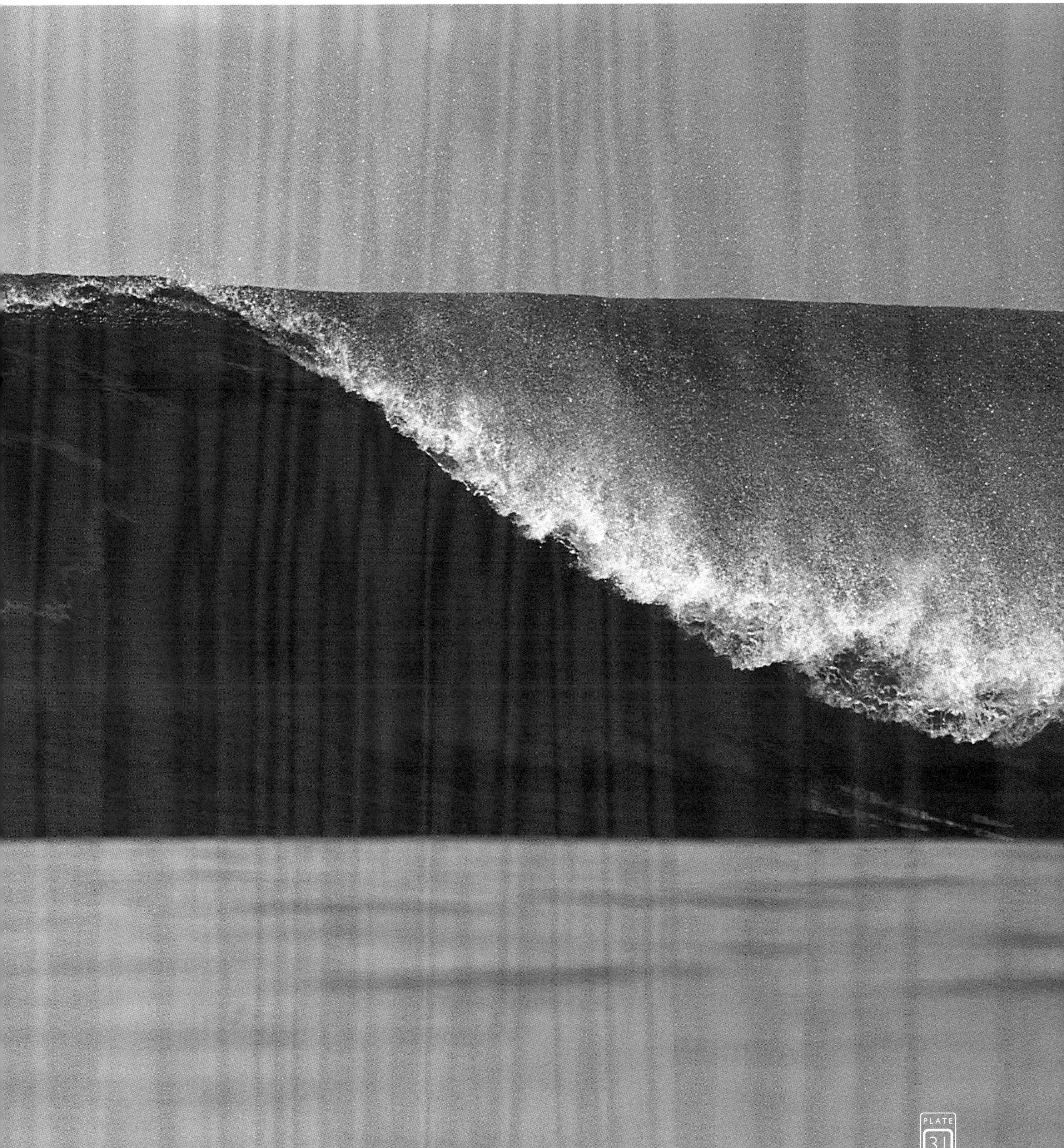

PLATE 31

Like as the waves make toward the pebbled shore,
So do our minutes hasten to their end.

William SHAKESPEARE SONNETS

1592 – 1598

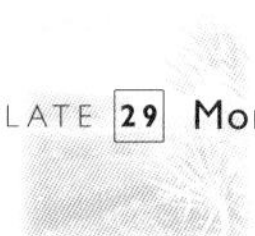

PLATE 29 Morning waves peel down the reef at Hanalei Bay, Hawaii. These islands have long been known as surfing's Mecca, thanks to the combination of big Pacific groundswells, favourable tradewinds, the lack of a swell-slowing continental shelf, and the long lava-flow reefs set at often ideal angles for the approaching waves.

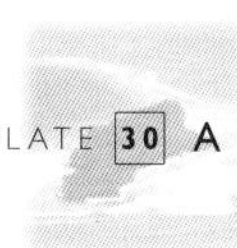

PLATE 30 A classic example of refraction, as long-range Indian Ocean groundswells flex and ricochet around a small coral island in the southern Maldives. Refraction is caused by the slowing of part of a swell as it moves across a shallow bottom contour; it bends toward the shallow in a sweeping arc, forming beautiful surfing waves.

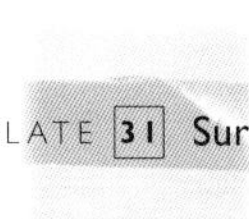

PLATE 31 Surfer's-eye view of a hollow peeling wave in Samoa. Tropical South Pacific islands are constantly exposed to powerful swell action; in fact, it was in this region, during the U.S. World War II invasion effort of Japanese-held islands, that the modern science of wave forecasting was invented, in order to prevent troop carriers being trapped in a surf zone.

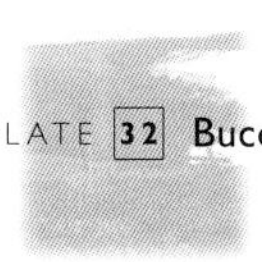

PLATE 32 Bucolic scenery frames a dramatic coastal change at Crescent Head, NSW. On the open side of the point, harsh serrations of ancient volcanic rock have been exposed by heavy wave action; within the shelter of the bay further around, a smooth sandy curve of beach accumulates. Coastlines are happy hunting grounds for geologists, for whom the waves expose millions of years of planetary history.

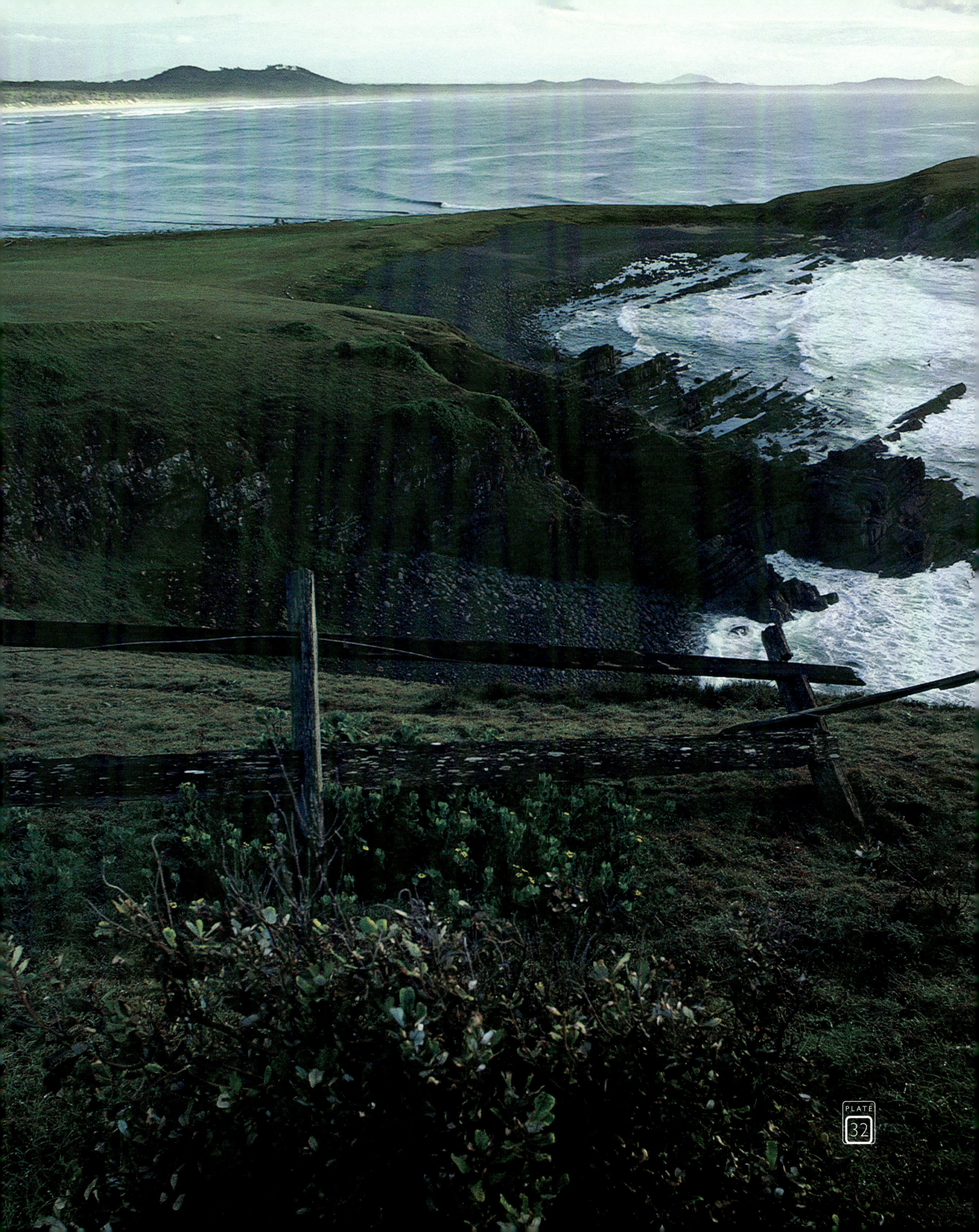

PLATE 32

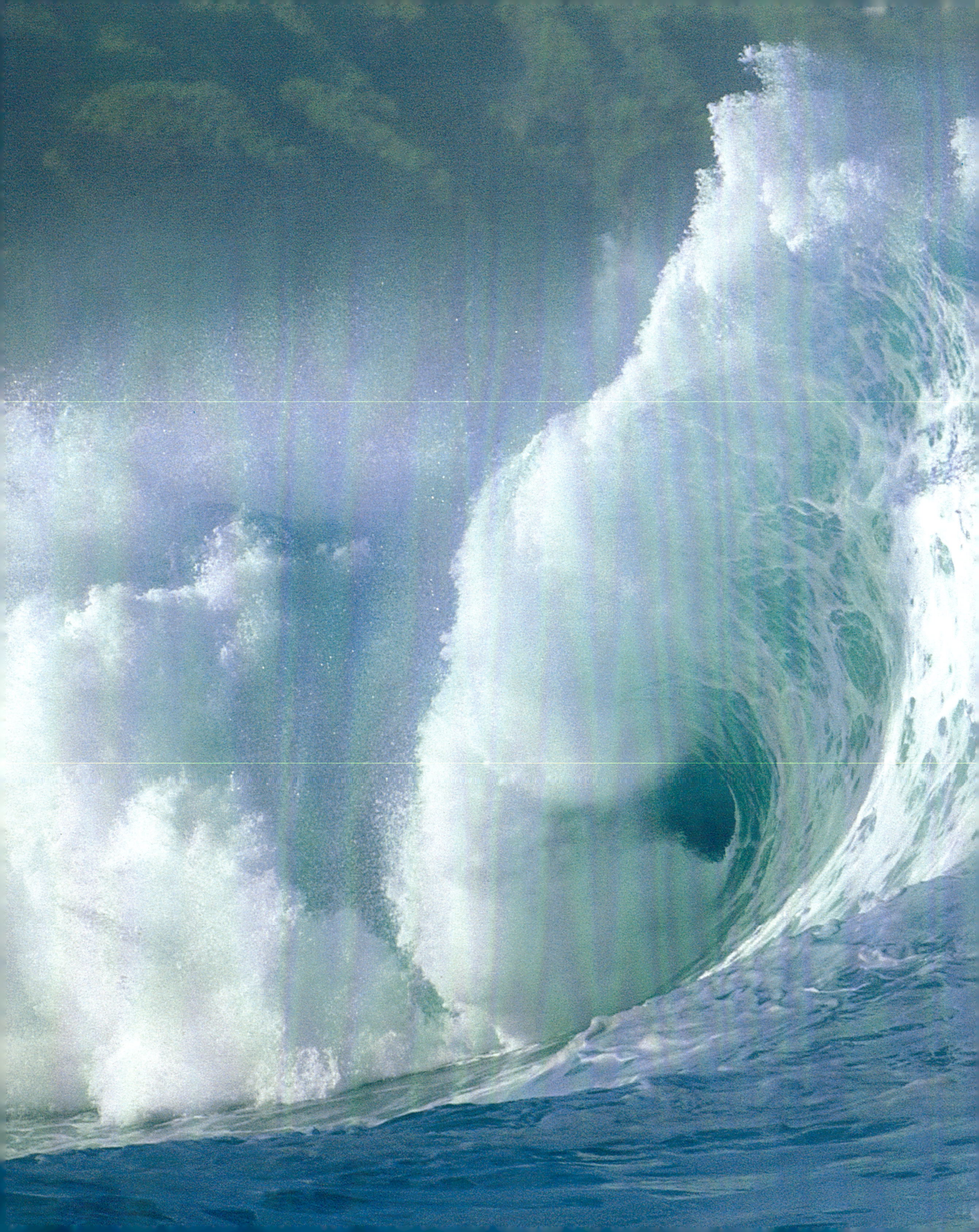

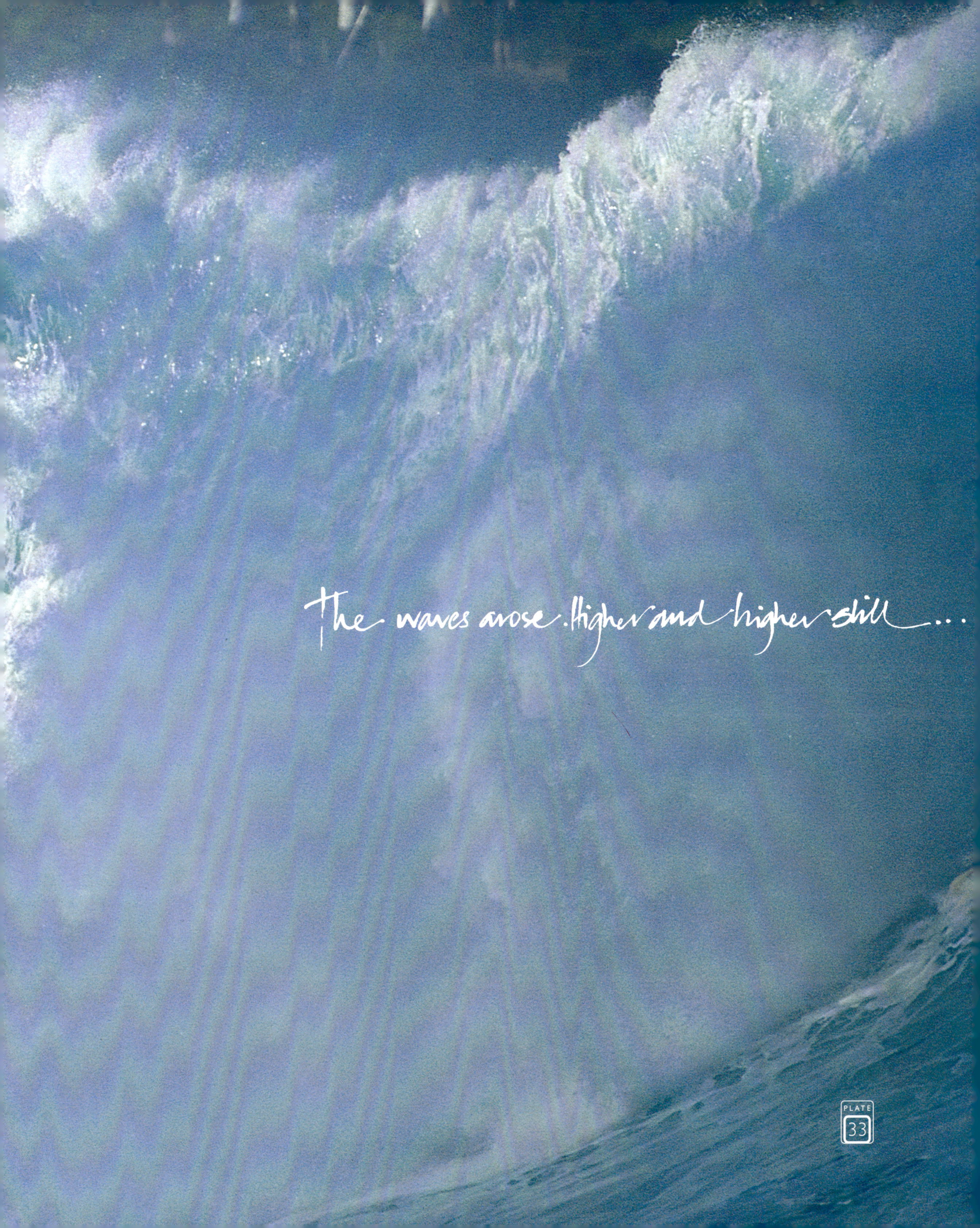
The waves arose. Higher and higher still . . .
PLATE
33

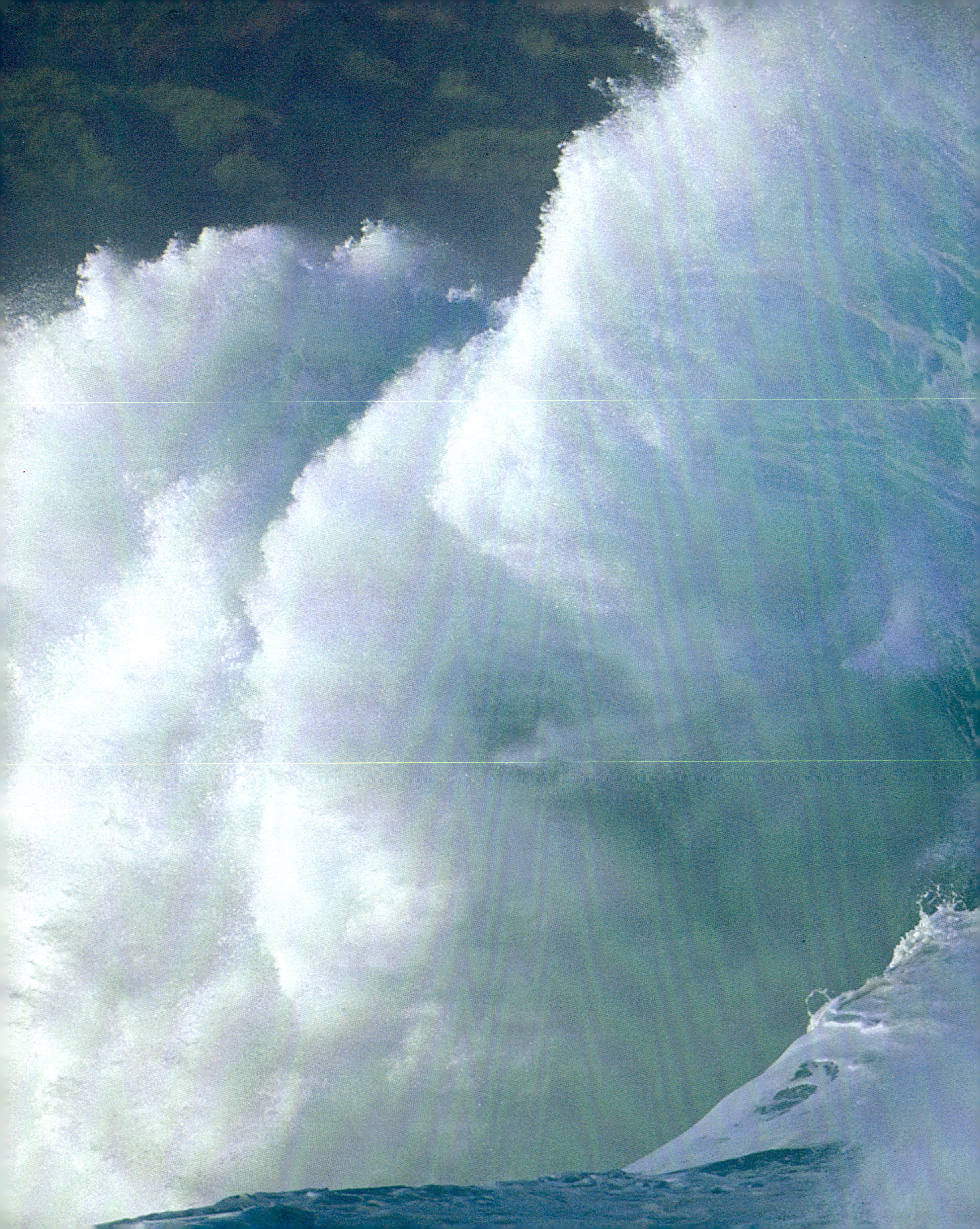

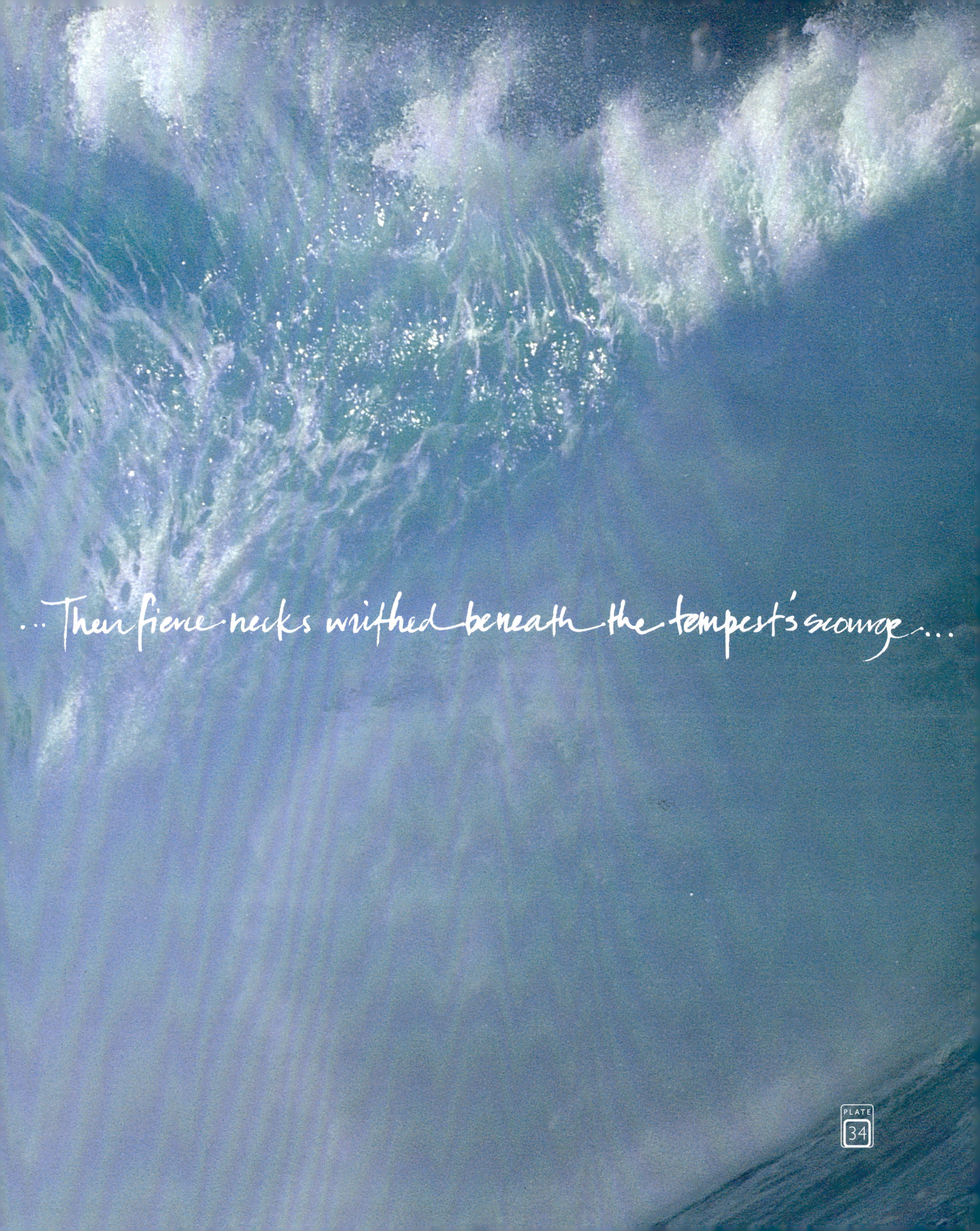

... Their fierce necks writhed beneath the tempest's scourge ...

PLATE 34

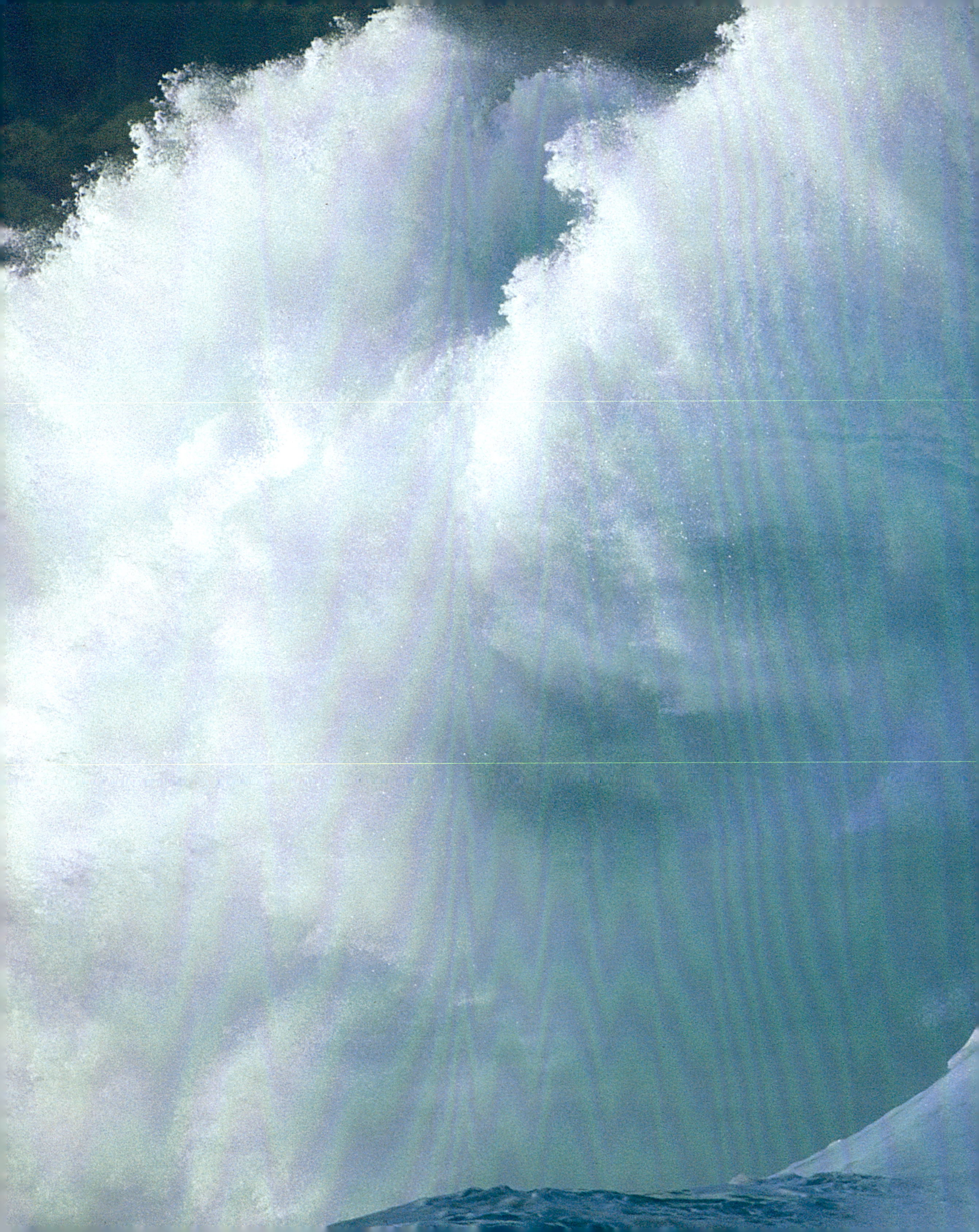

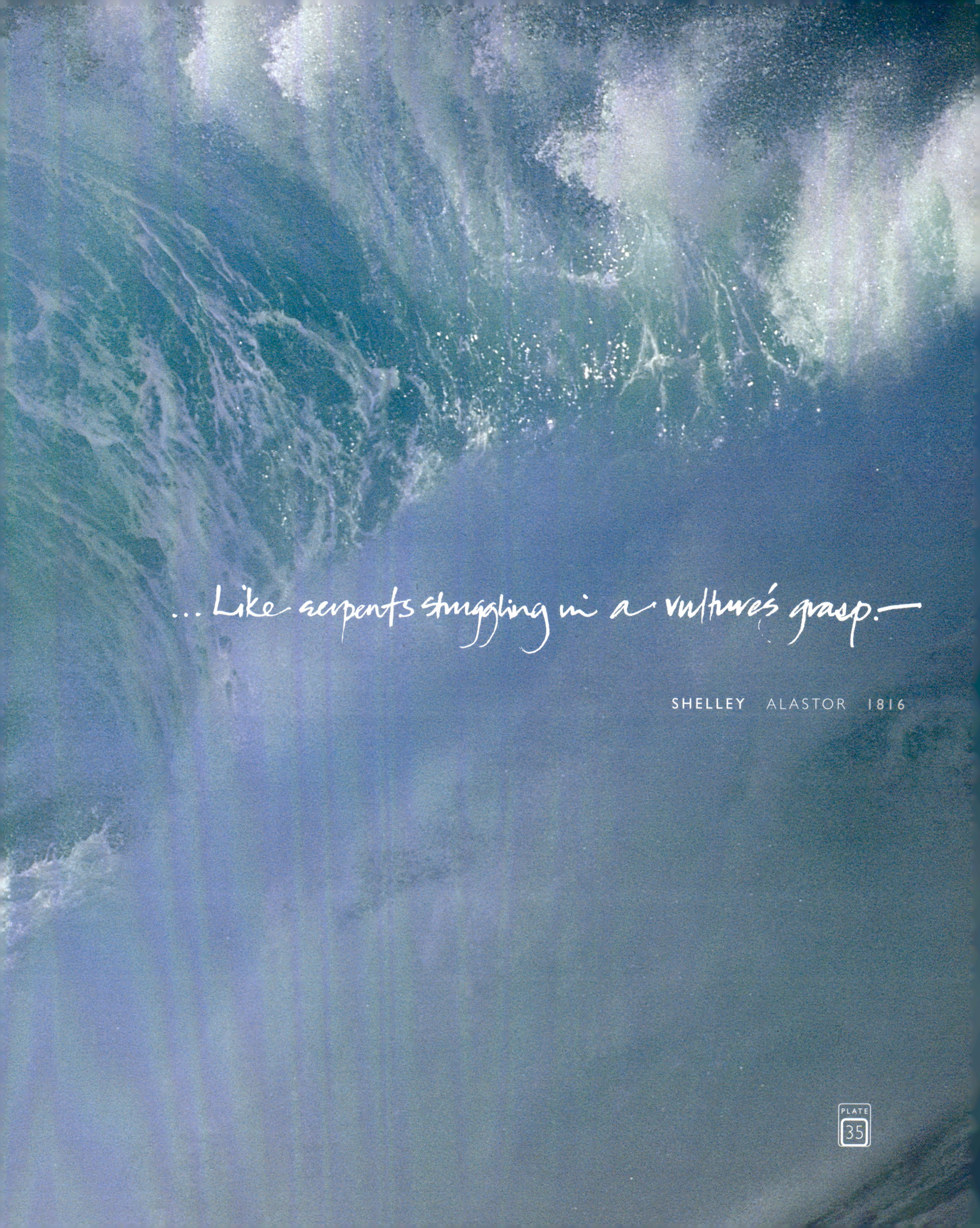

... Like serpents struggling in a vulture's grasp.—

SHELLEY ALASTOR 1816

O most powerful and glorious Lord God, at whose command the winds blow, and lift up the waves of the sea, and who stillest the rage thereof; We thy creatures, but miserable sinners, do in this our great distress cry unto thee for help; Save, Lord, or else we perish. We confess, when we have been safe and seen all things quiet about us, we have forgot thee our God, and refused to hearken to the still voice of thy word, and to obey thy commandments: But now we see how terrible thou art in all thy works of wonder; the great God to be feared above all: and therefore we adore thy divine Majesty, acknowledging thy power, and imploring thy goodness.

PRAYERS TO BE USED IN STORMS AT SEA BOOK OF COMMON PRAYER 1662

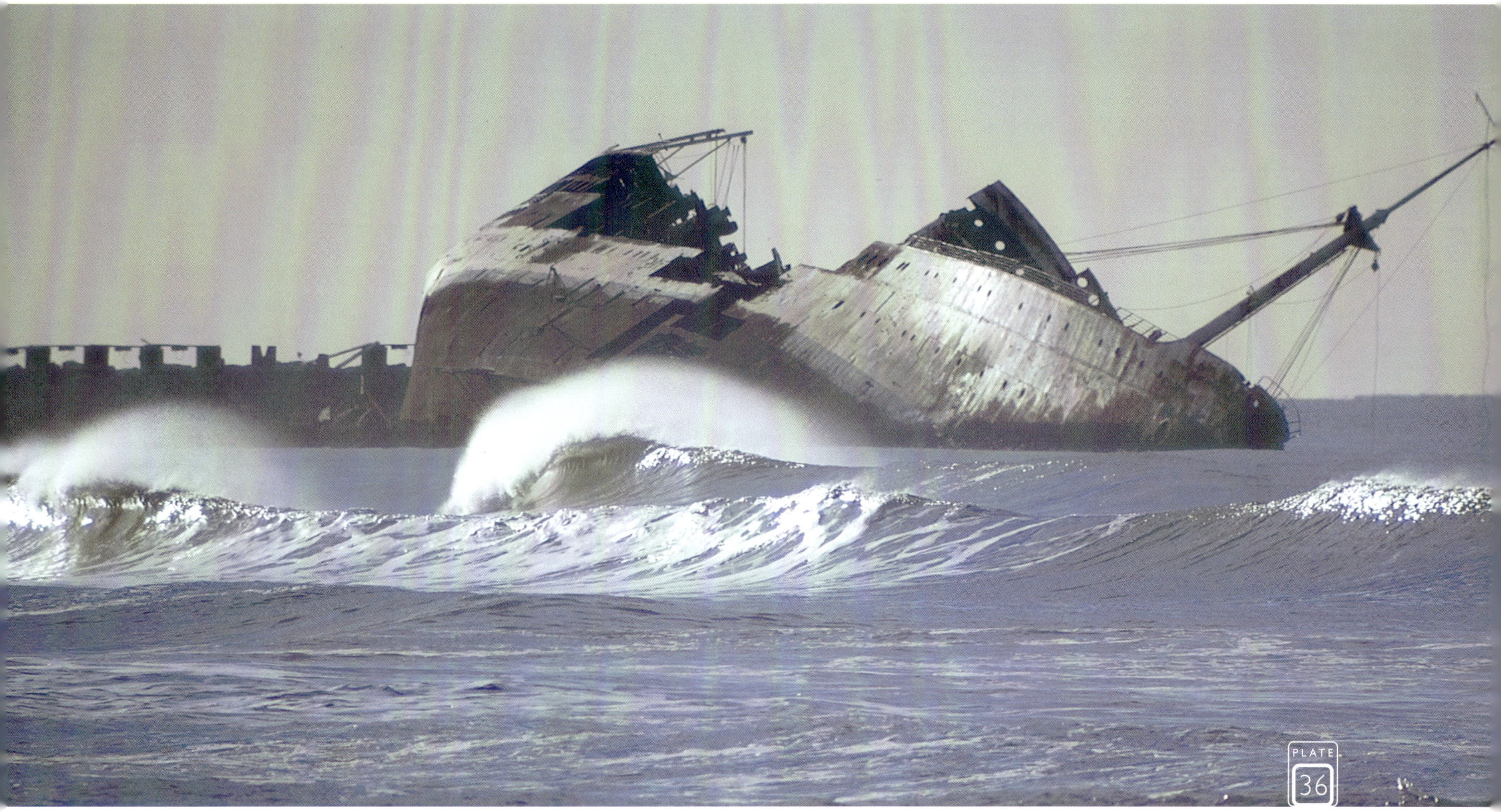
PLATE
36

I have seen the hungry ocean gain
Advantage on the kingdom of the shore,
And the firm soil win of the watery main,
Increasing store with loss and loss with store.

William SHAKESPEARE SONNETS

1592 – 1598

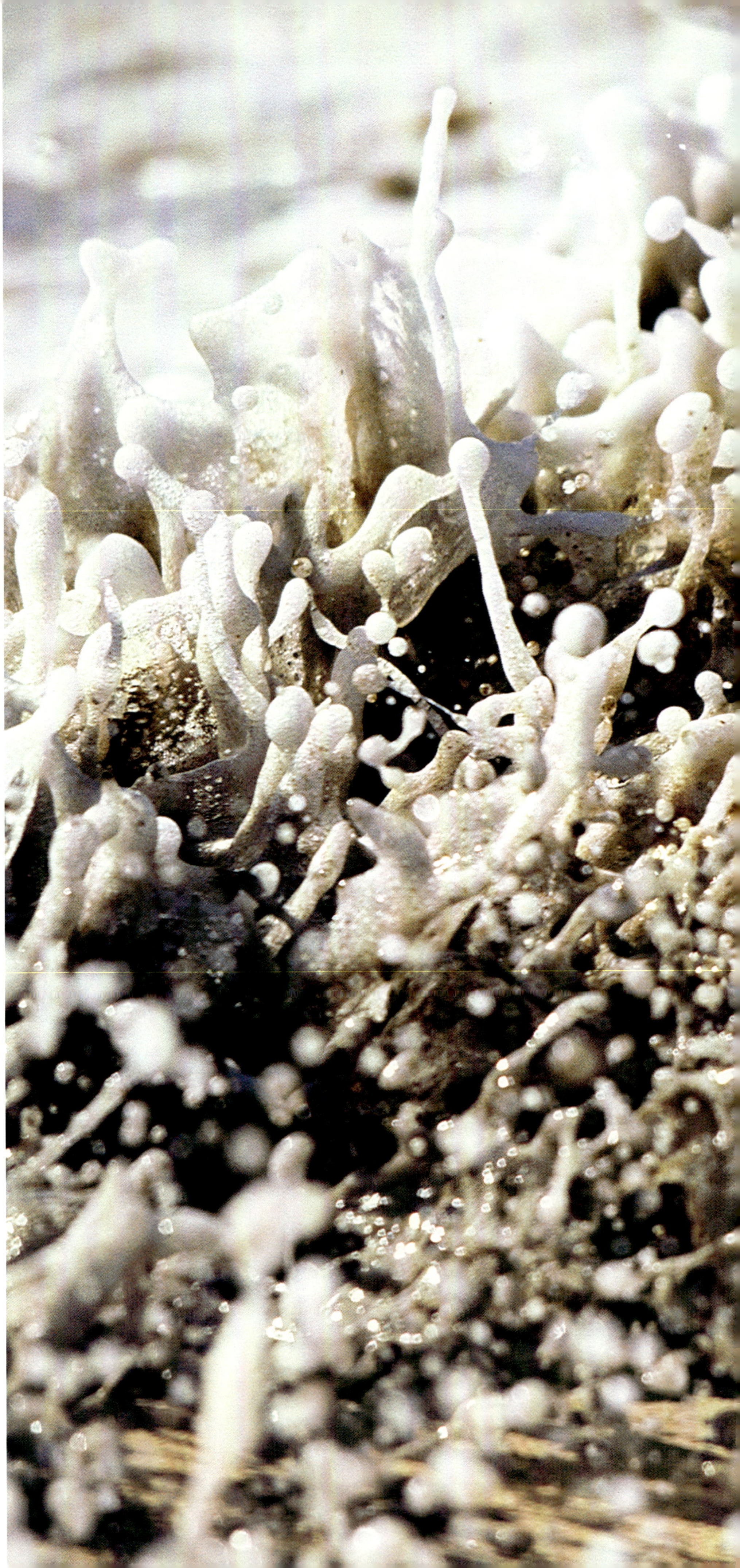

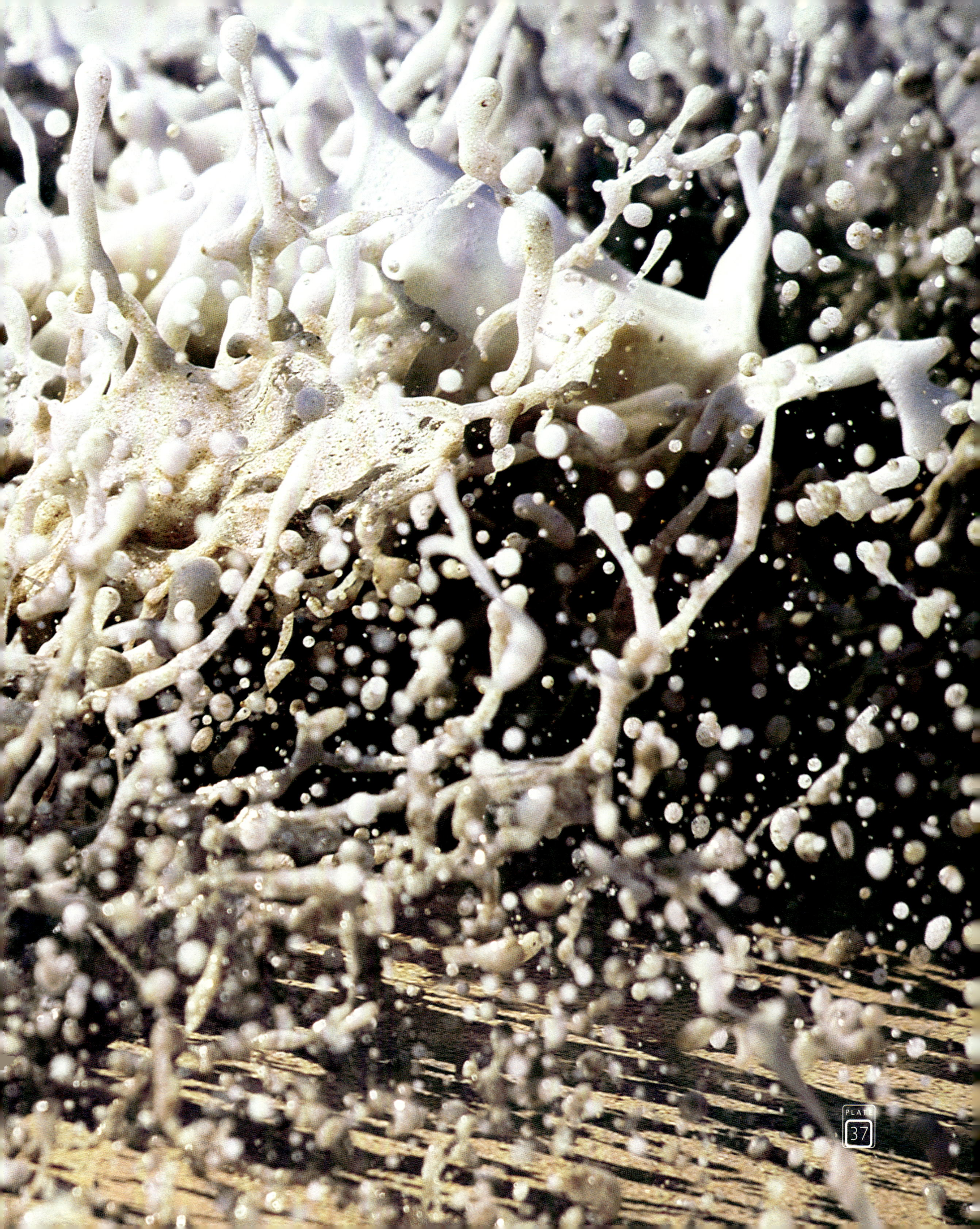

PLATE 37

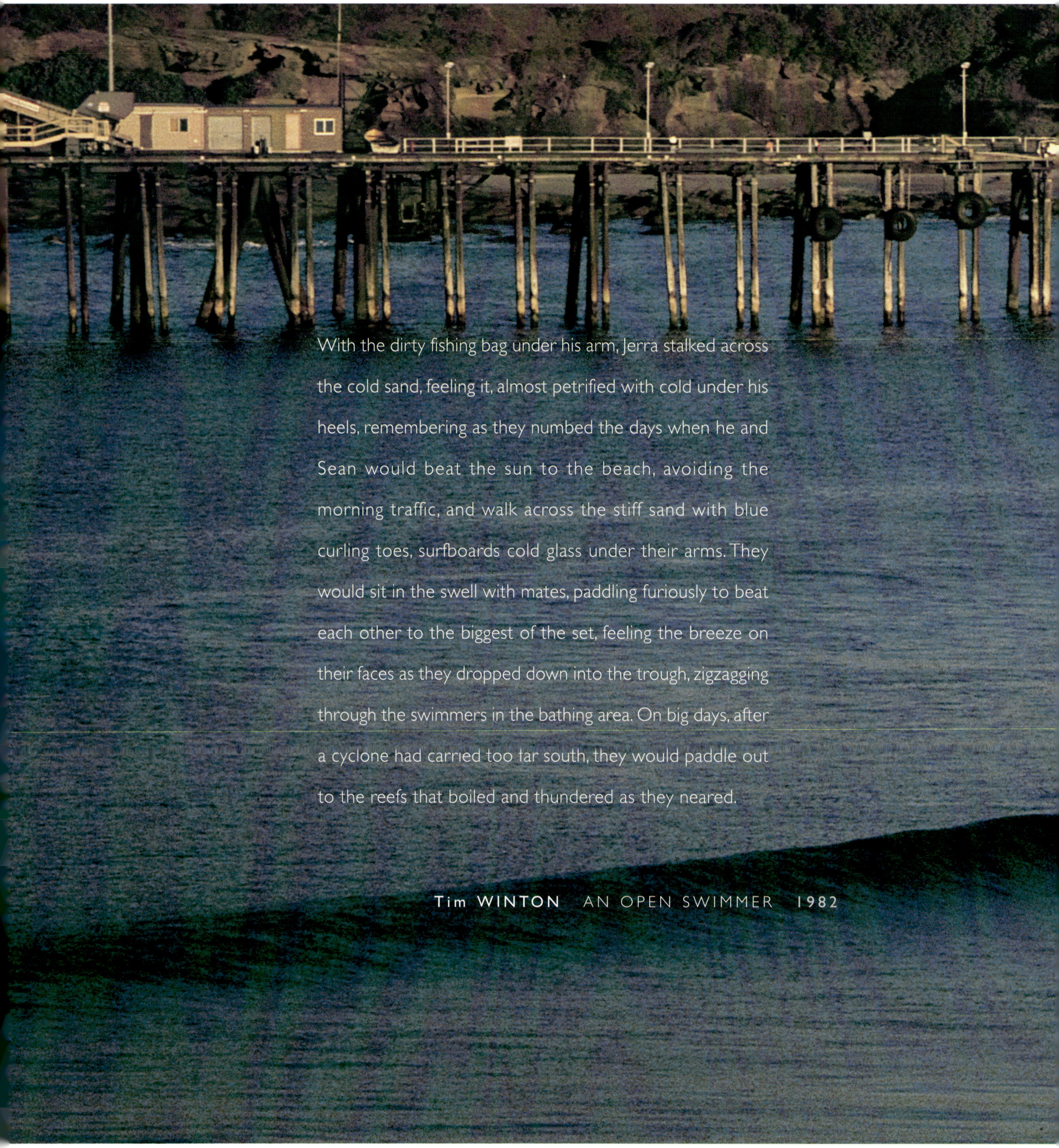

With the dirty fishing bag under his arm, Jerra stalked across the cold sand, feeling it, almost petrified with cold under his heels, remembering as they numbed the days when he and Sean would beat the sun to the beach, avoiding the morning traffic, and walk across the stiff sand with blue curling toes, surfboards cold glass under their arms. They would sit in the swell with mates, paddling furiously to beat each other to the biggest of the set, feeling the breeze on their faces as they dropped down into the trough, zigzagging through the swimmers in the bathing area. On big days, after a cyclone had carried too far south, they would paddle out to the reefs that boiled and thundered as they neared.

Tim WINTON AN OPEN SWIMMER 1982

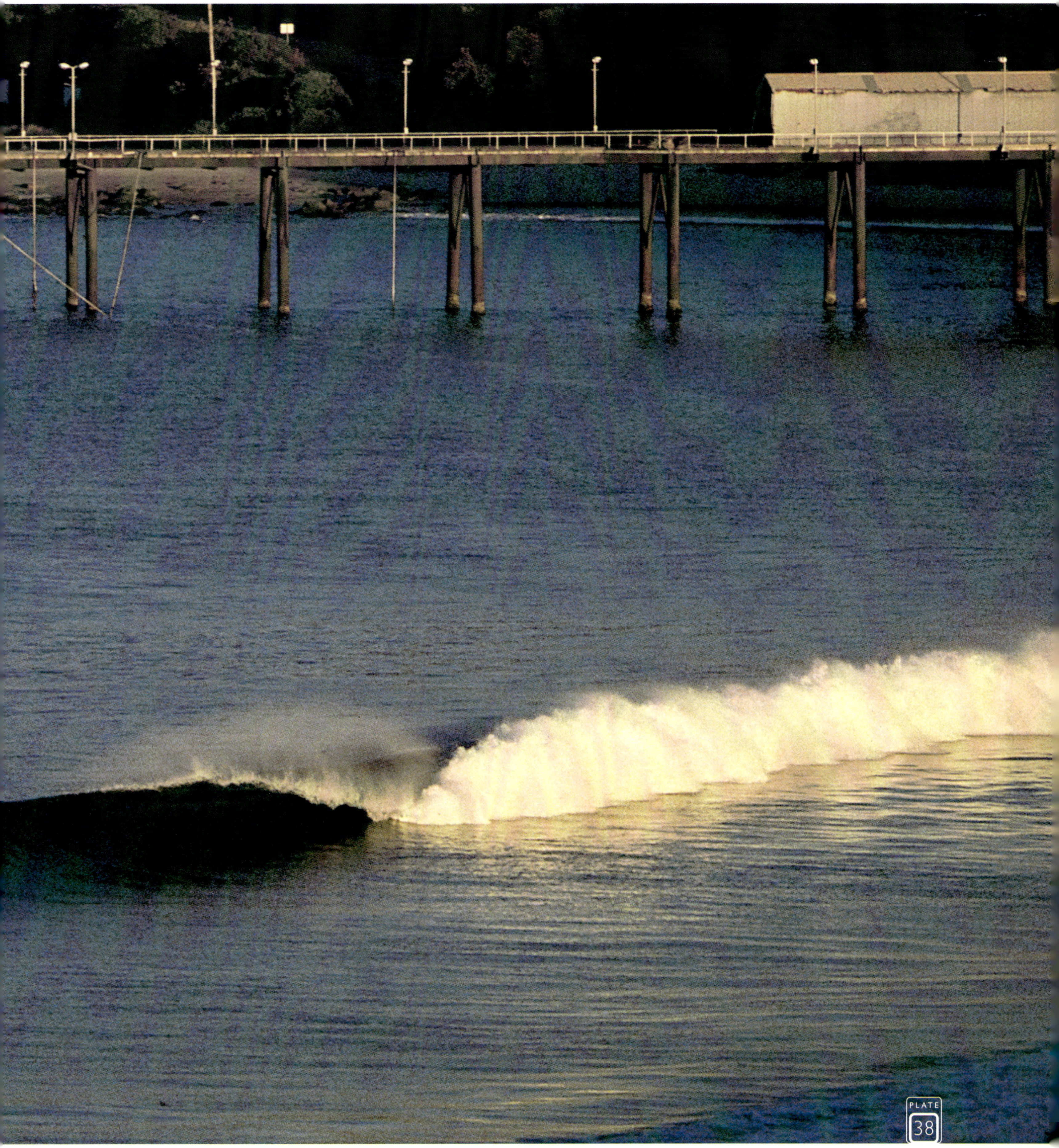

PLATE
38

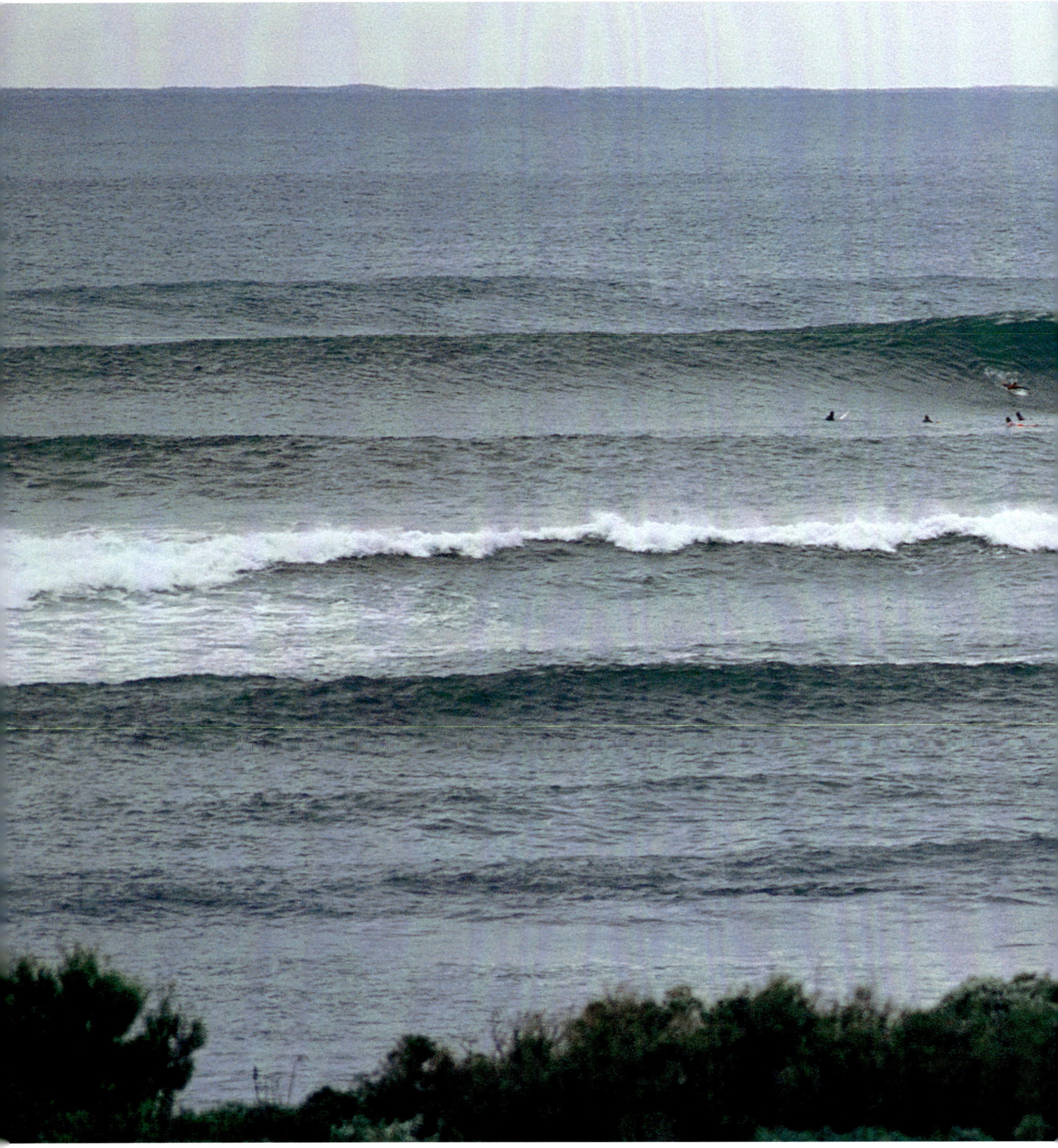

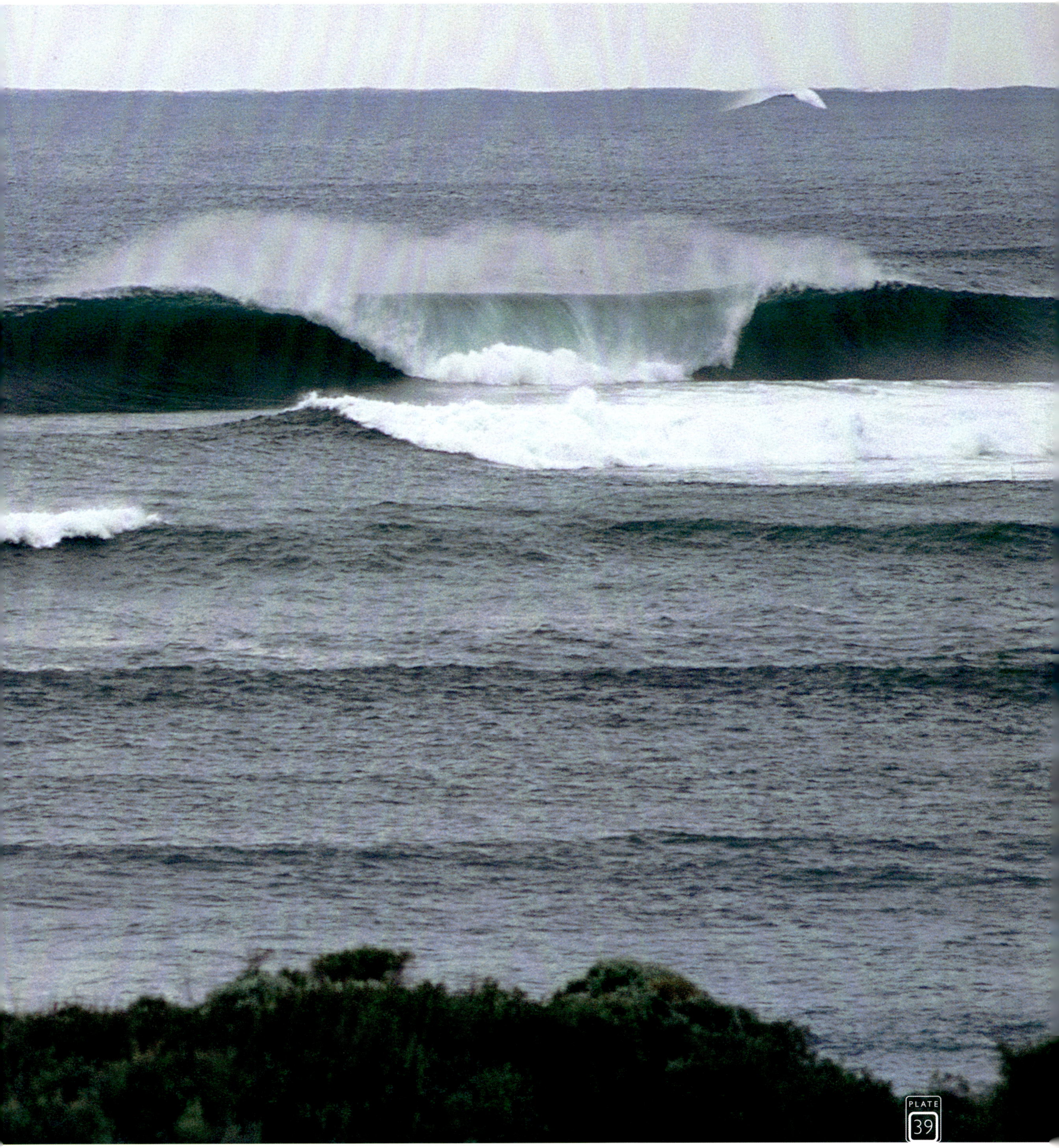

PLATE 39

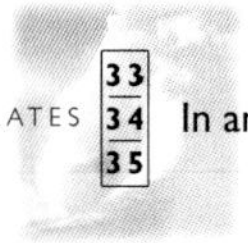

PLATES 33 34 35 In an expression of the North Pacific's outrageous power, a massive shorebreak literally explodes at Oahu's Waimea Bay. To the onlooker, it appears impossible that the human frame could survive such a wave's impact — and indeed, being caught in the Waimea shorebreak's violent grip is surely the stuff of nightmares.

PLATE 36 Playing innocent, some small waves dance harmlessly in the shadow of their big brothers' victim — a shattered cargo vessel. Even the biggest supertankers still run risks from rogue deep ocean swells, some of which have been recorded at over 100 feet. Even more dangerous are troughs caused by the collision of big swells and opposing ocean currents; ships coming upon them unawares can be crippled in moments.

PLATE 37 Like a human army advancing on its foe, a small yet powerful wave erupts onto a sandy beach. Wave action such as this is constantly at work on the coasts of the world, wearing shells and rock and coral into sand, and drawing the sand deep into the sea, where one day it might finally settle into the dusty silts of the abyssal plains, beyond the continental shelves.

The surf changes from moment to moment, day to day, and beach to beach. The waves are influenced by the bottom and the bottom is changed by the waves. And since the waves arriving at a beach are highly variable in height, period, and direction, each wave creates a slightly different bottom configuration for the ones that come after it. The water level changes with the tide and the waves change as the storms at sea develop, shift position, and die out again. The result is that sand bottom is forever being rearranged. Even in glass-sided wave channels where an endless number of waves, each exactly the same, can be produced, equilibrium is never reached; the sand continues to change as long as the wave machine is running.

Williard BASCOM WAVES AND BEACHES 1958

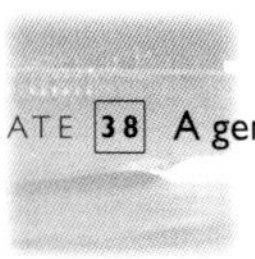

PLATE 38 A gentle wave peels across a quiet sandbar at Catherine Hill Bay, central NSW. Sheltered bays like this provide windless sanctuary for surfers in search of a cleanly formed wave up and down the Australian east coast. The pier in the background once carried coal from nearby mines to ships for transport to smelters at Newcastle, just a few kilometres north.

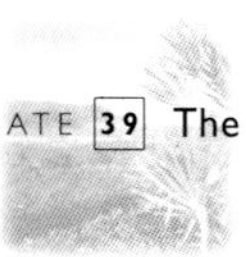

PLATE 39 The tremendous force of a deep Indian Ocean swell keeps surfers at more than arm's length in Cowaramup Bay, south-west Australia. Once spent on the outer reef, the swells in the foreground make their ways — reduced in power — to the beach. In such a broad bay, waves may break two or three times before finally expiring on the shore.

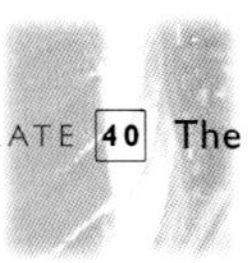

PLATE 40 The serrated knife-edge curve of a new swell defines the pointbreak at Rincon, California. Along with the refraction effect, that very human attraction to waves and coastlines is obvious — a highway hugs the beach, and million-dollar houses crowd each other for the ephemeral ocean view.

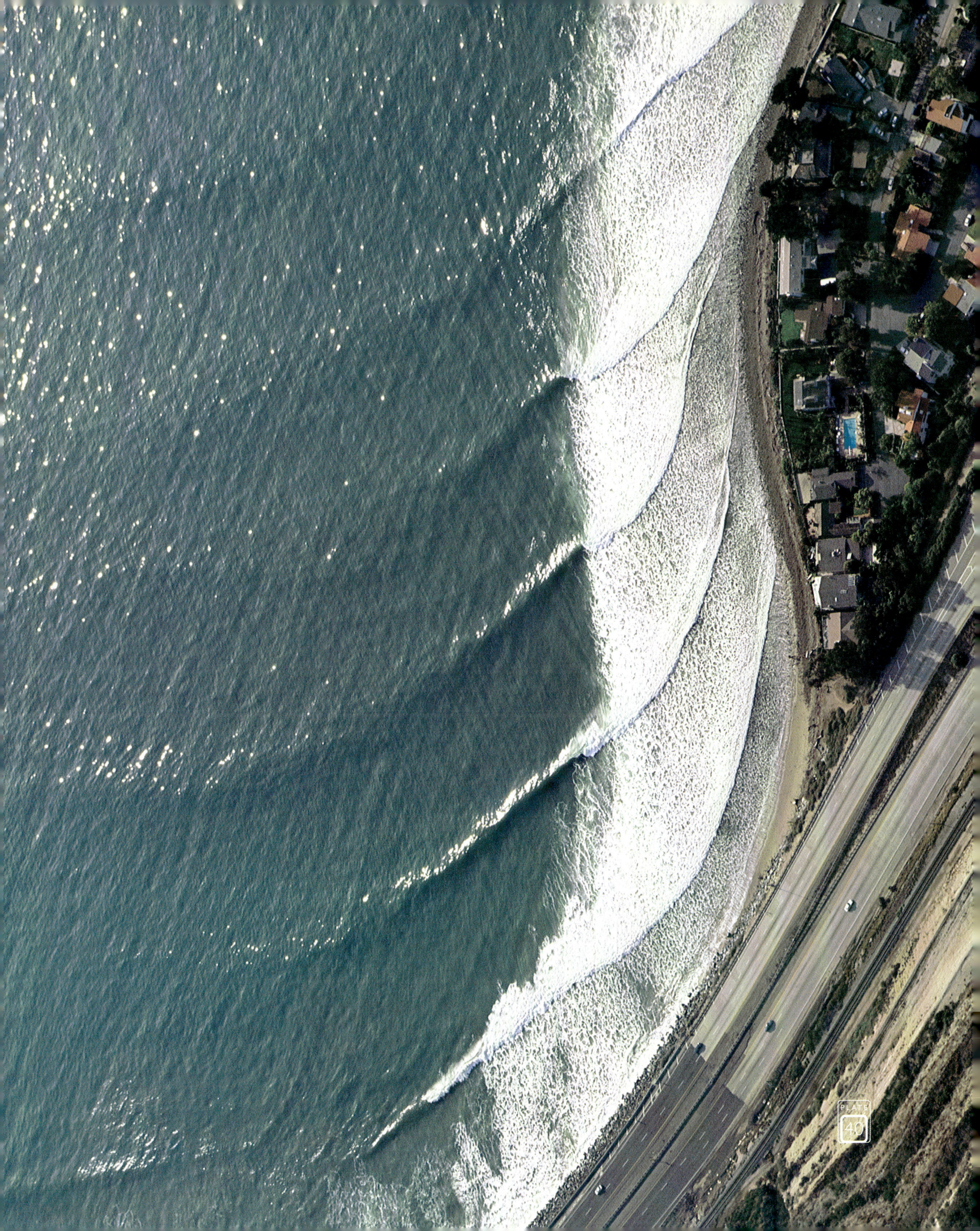
PLATE
40

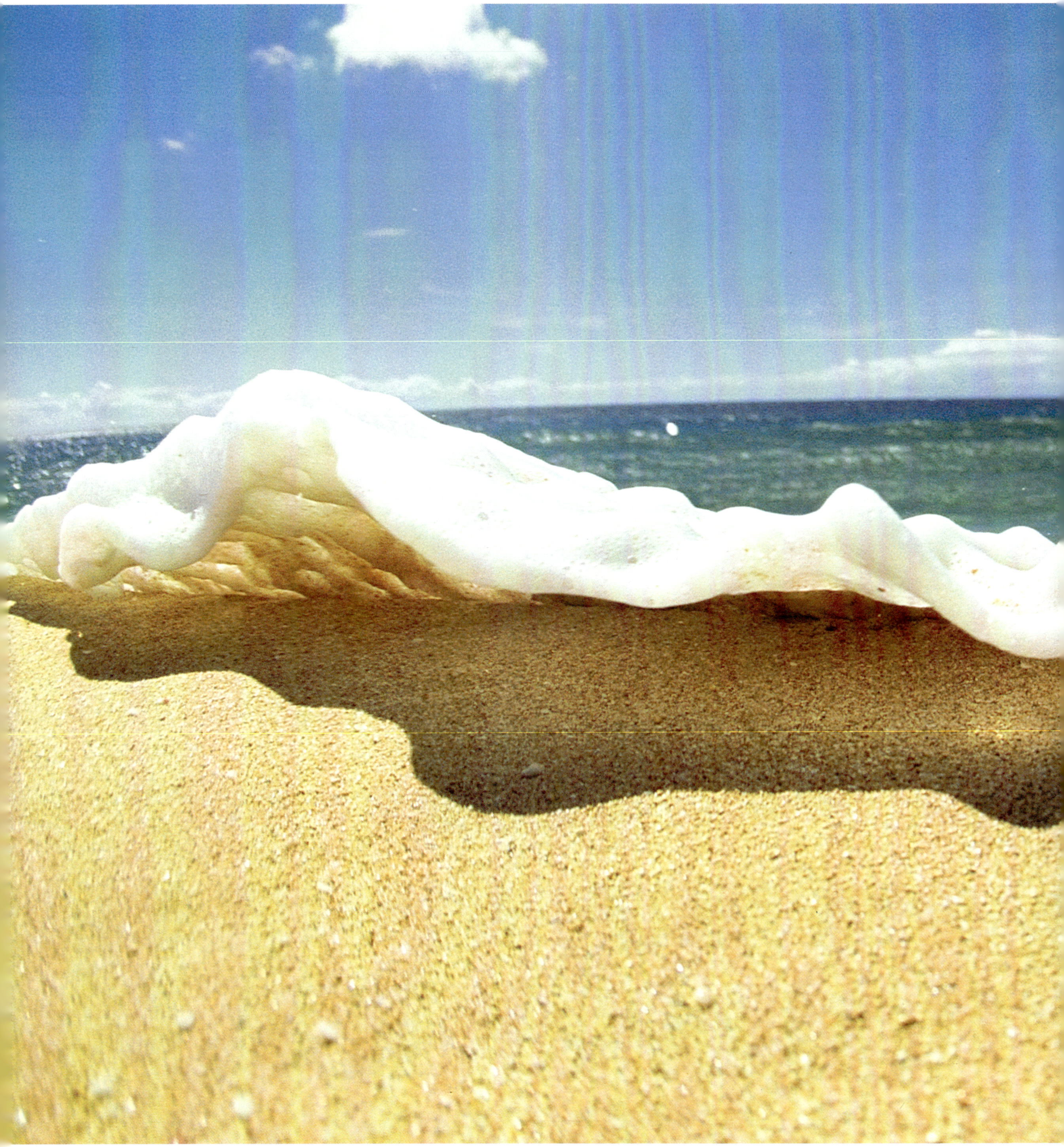

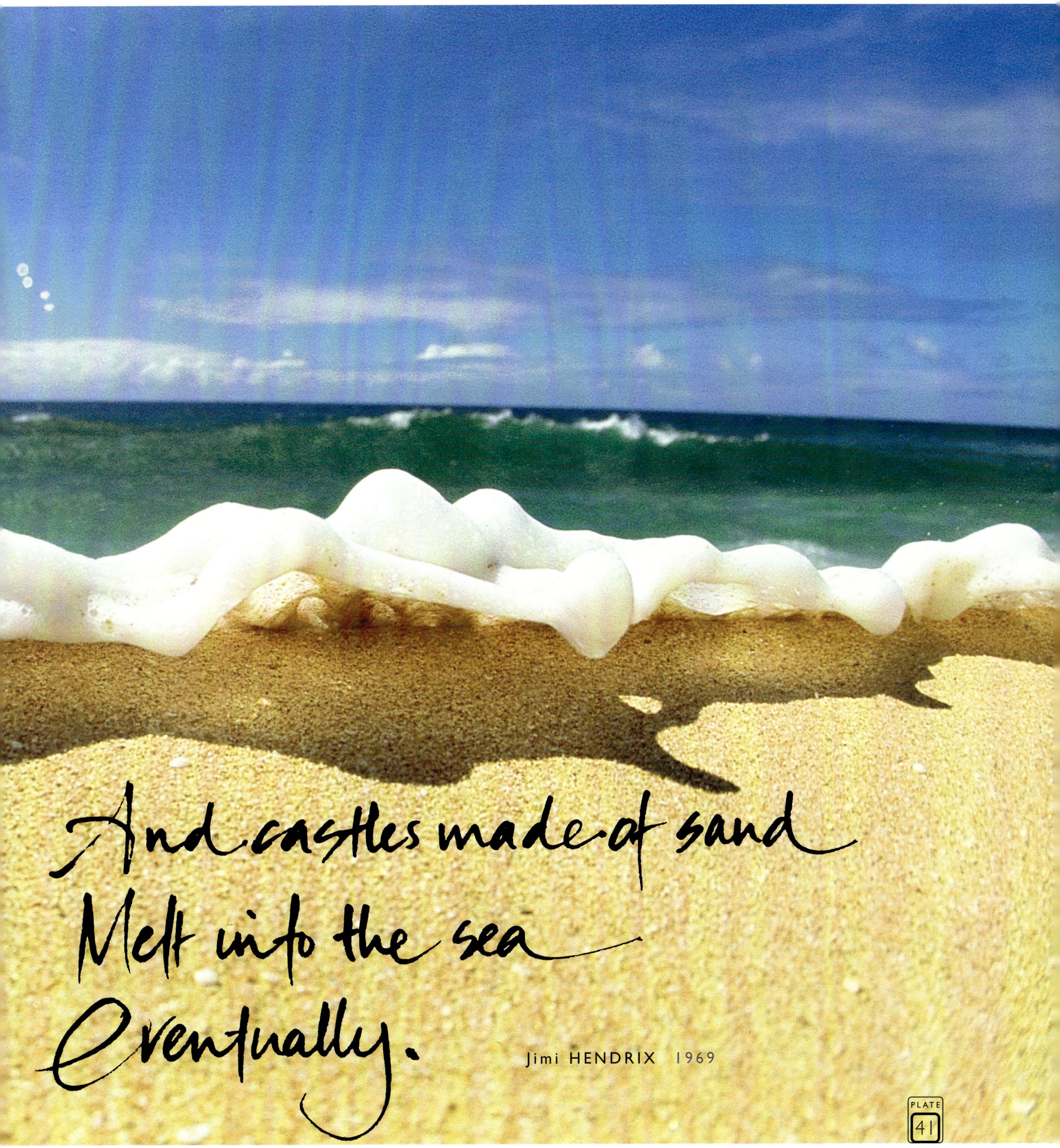
And castles made of sand
Melt into the sea
Eventually.
Jimi HENDRIX 1969
PLATE
41

An idea long existed that the ninth or tenth wave was greater and more powerful than others. This belief existed in Ovid's time, who says, 'The wave that is now coming o'ertops all the others; 'tis the one that comes after the ninth and before the tenth.' ... Allatius writes thus in 1645, during a voyage from Messina to Malta, 'I saw the captain, who was accounted an experienced and skilful mariner, standing at the bow, while he muttered and pointed at something with his finger. I approached, and inquired what he was doing. The old man, with a cheerful countenance, answered, "I am breaking the force of a fatal wave, and am making the sign of the cross, and saying the prayers proper for the occasion." I said, "Do you, then, know, amongst all these waves, which is the fatal one?" "Yes," he said, "by so many waves by which the ship is tossed, none but the ninth can sink it."'

LEGENDS AND SUPERSTITIONS OF THE SEA

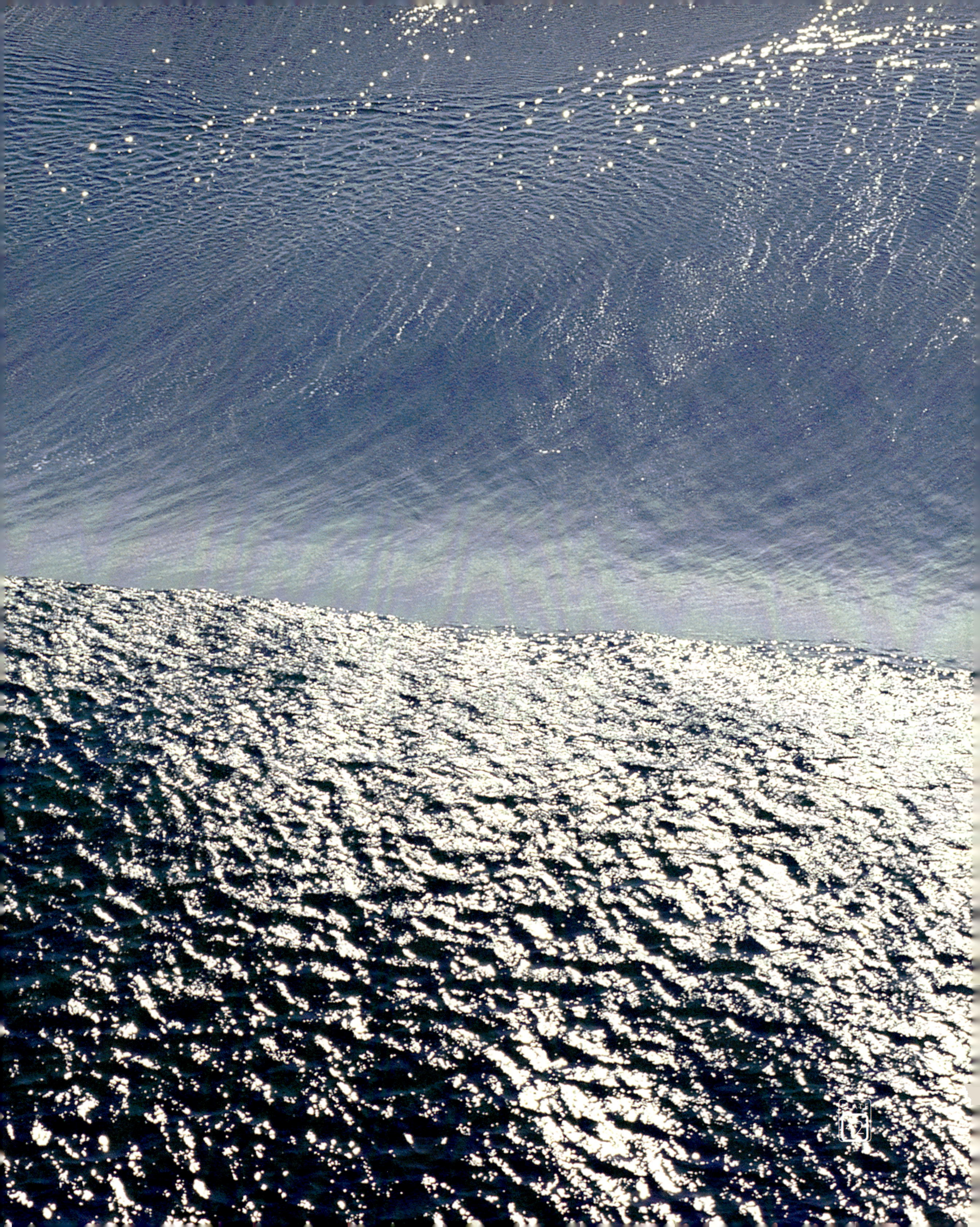

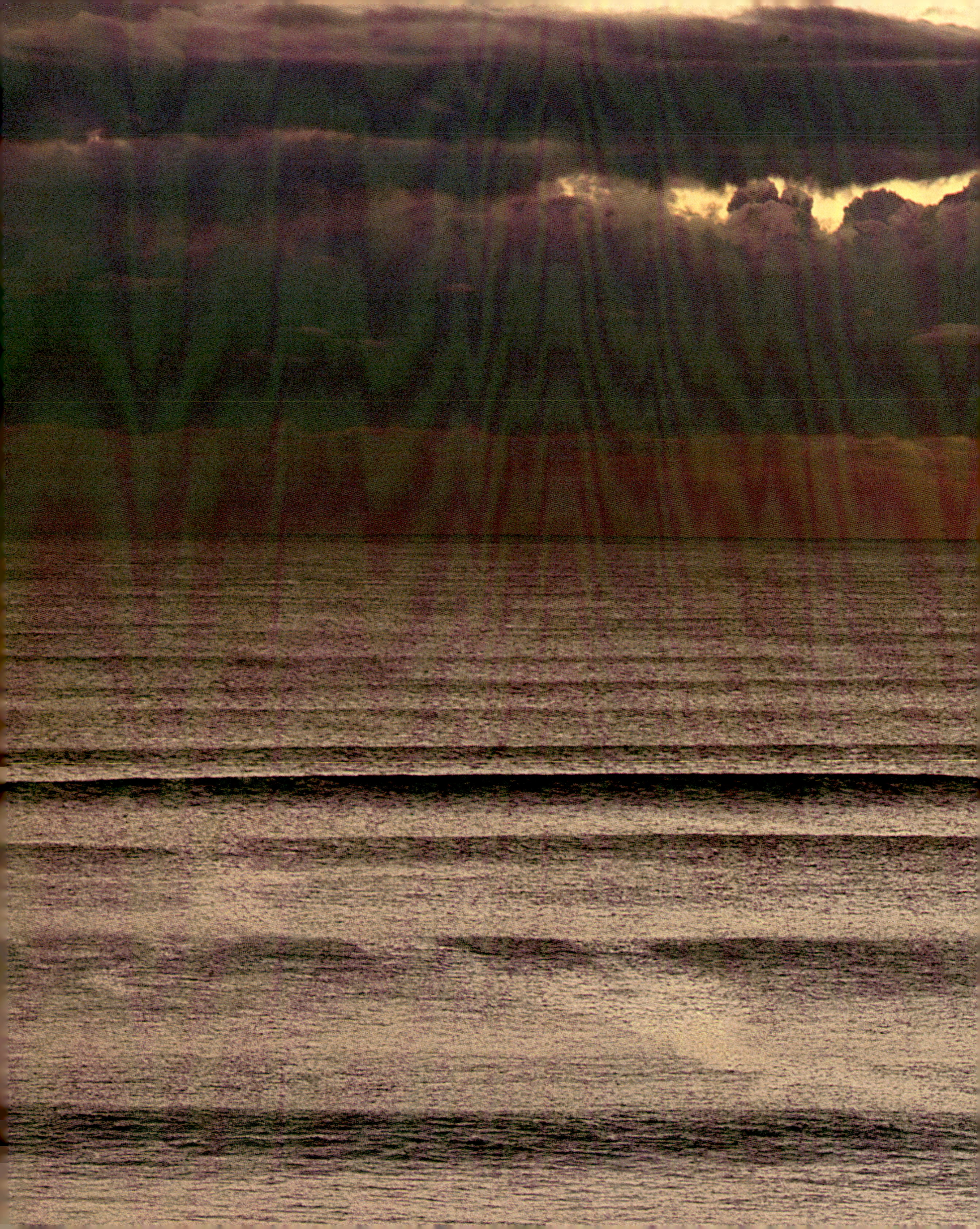

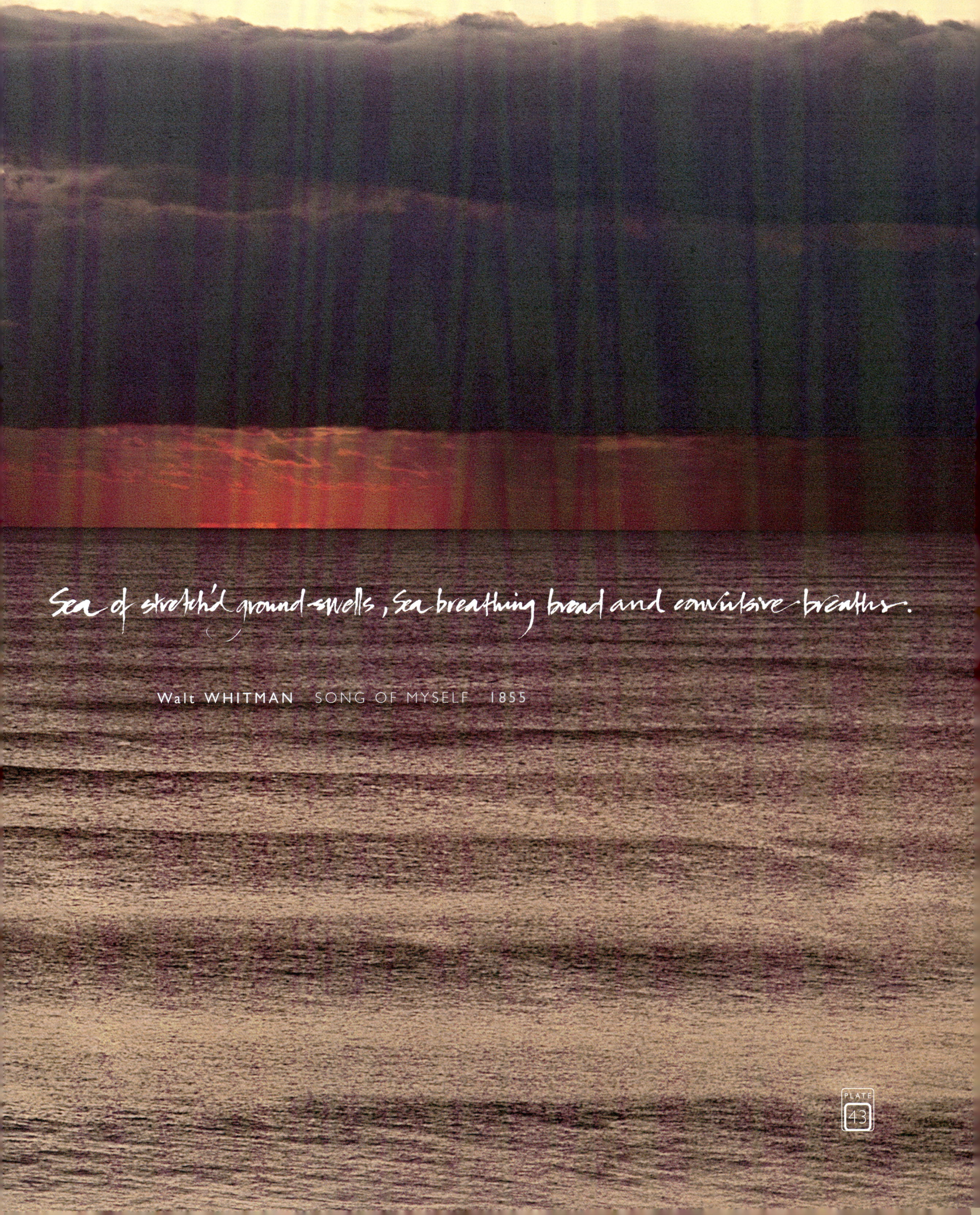
Sea of stretch'd ground-swells, Sea breathing broad and convulsive breaths.
Walt WHITMAN SONG OF MYSELF 1855
PLATE 43

If you travel in a boat toward oncoming water waves, the waves appear to approach you quickly. As you ride over these waves, they appear steep and frequent. But if you turn the boat around you can travel in the same direction and at the same speed as the wave so that, relative to you, it appears not to be moving at all. If you sailed just a little slower than the waves, you would feel your boat gently rise and fall as the waves passed beneath you. At a very special sailing speed, the wave speed appears to be zero and the wavelength infinite. You would continually sail down the wave which seems to go on forever. This is exactly the principle of surfing.

David BLAIR + Geoff McNAMARA RIPPLES ON A COSMIC SEA 1998

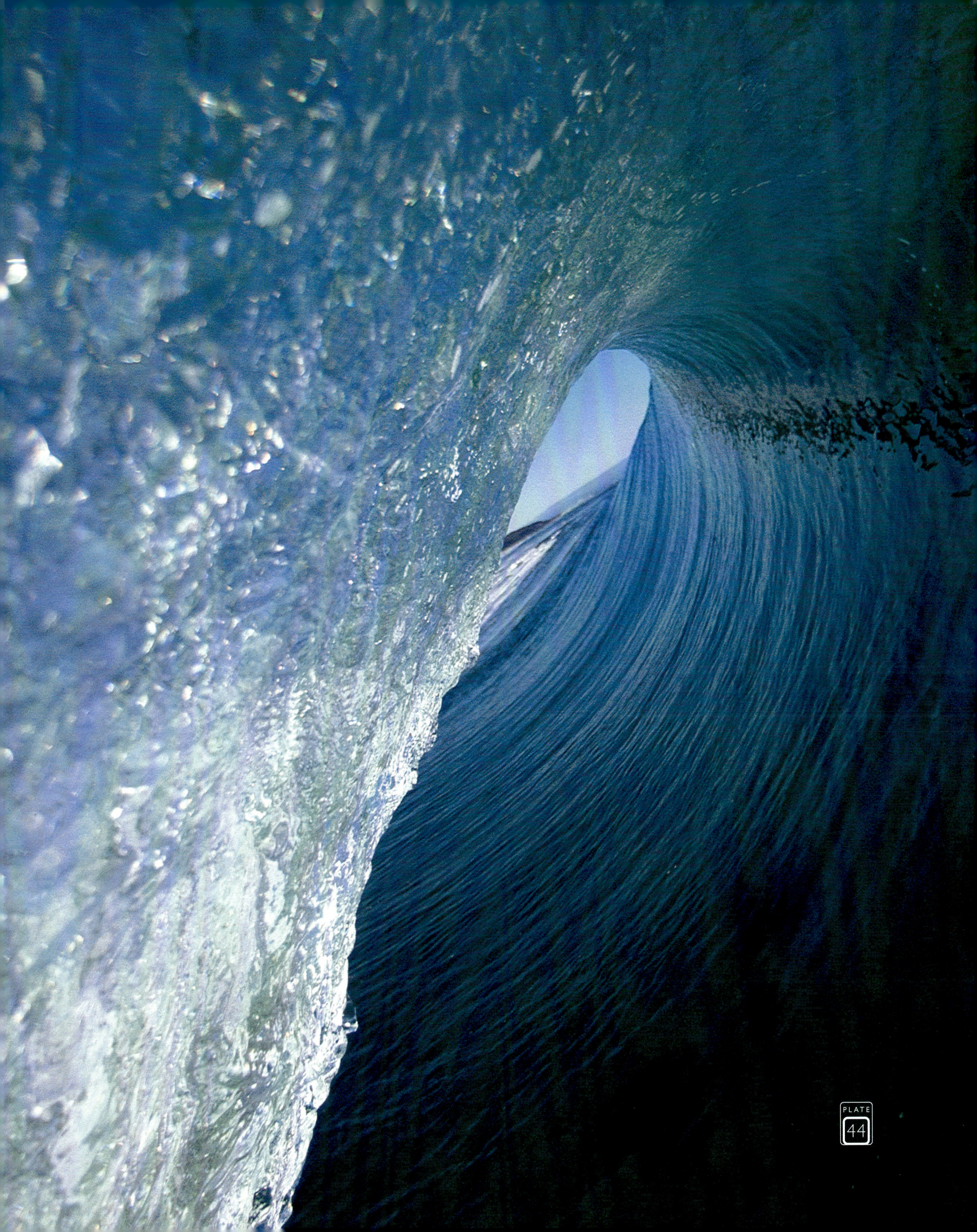

PLATE 44

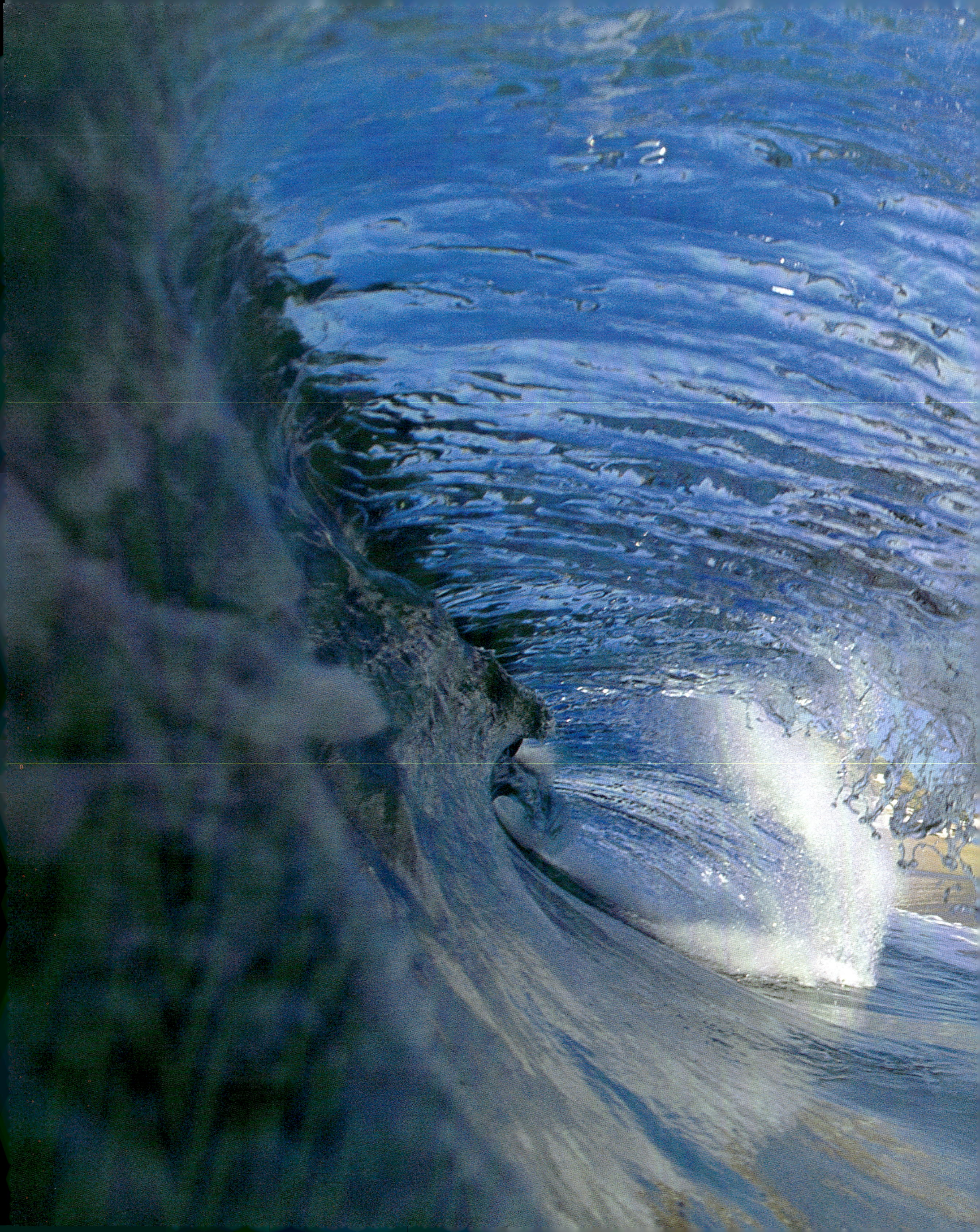

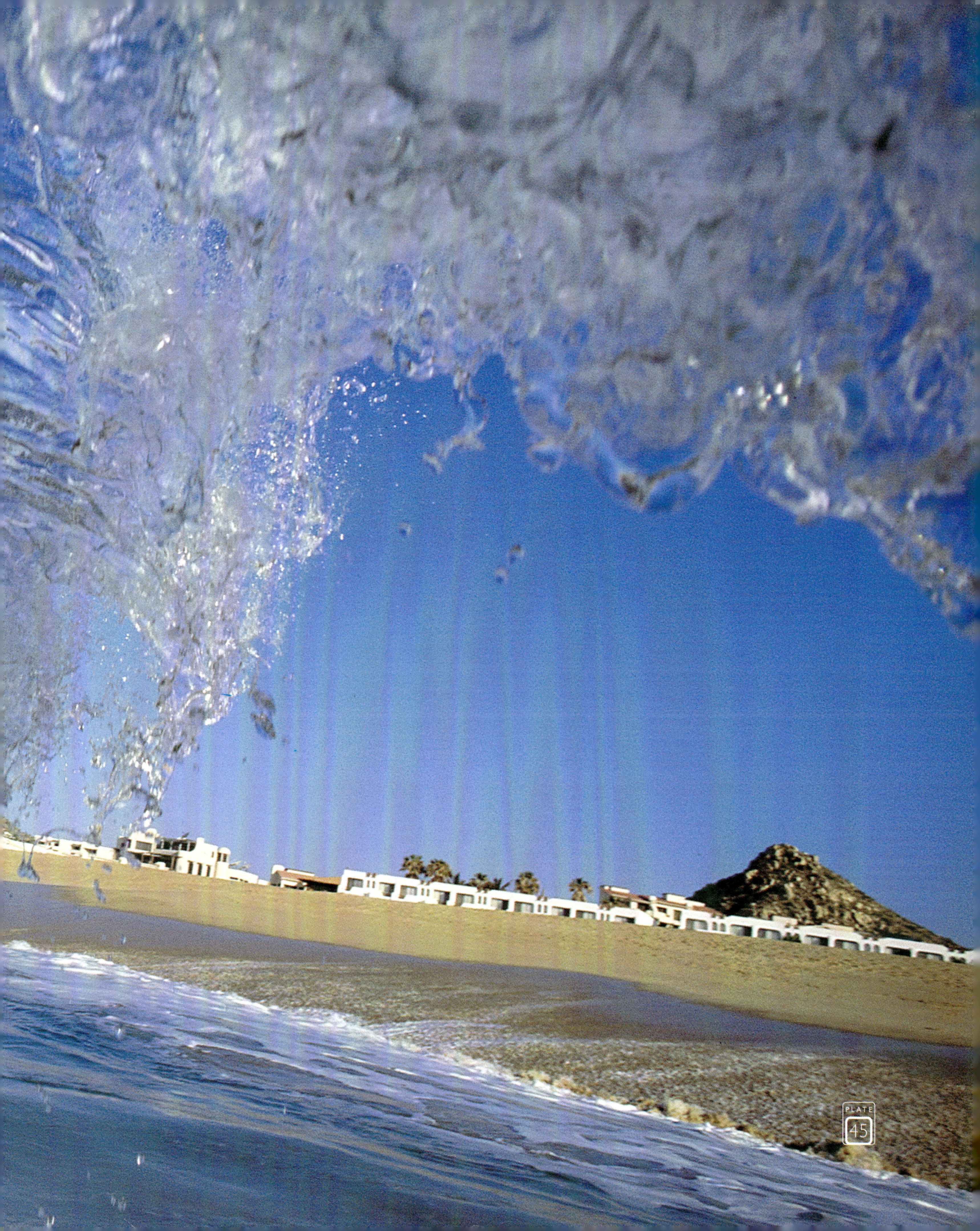

PLATE 45

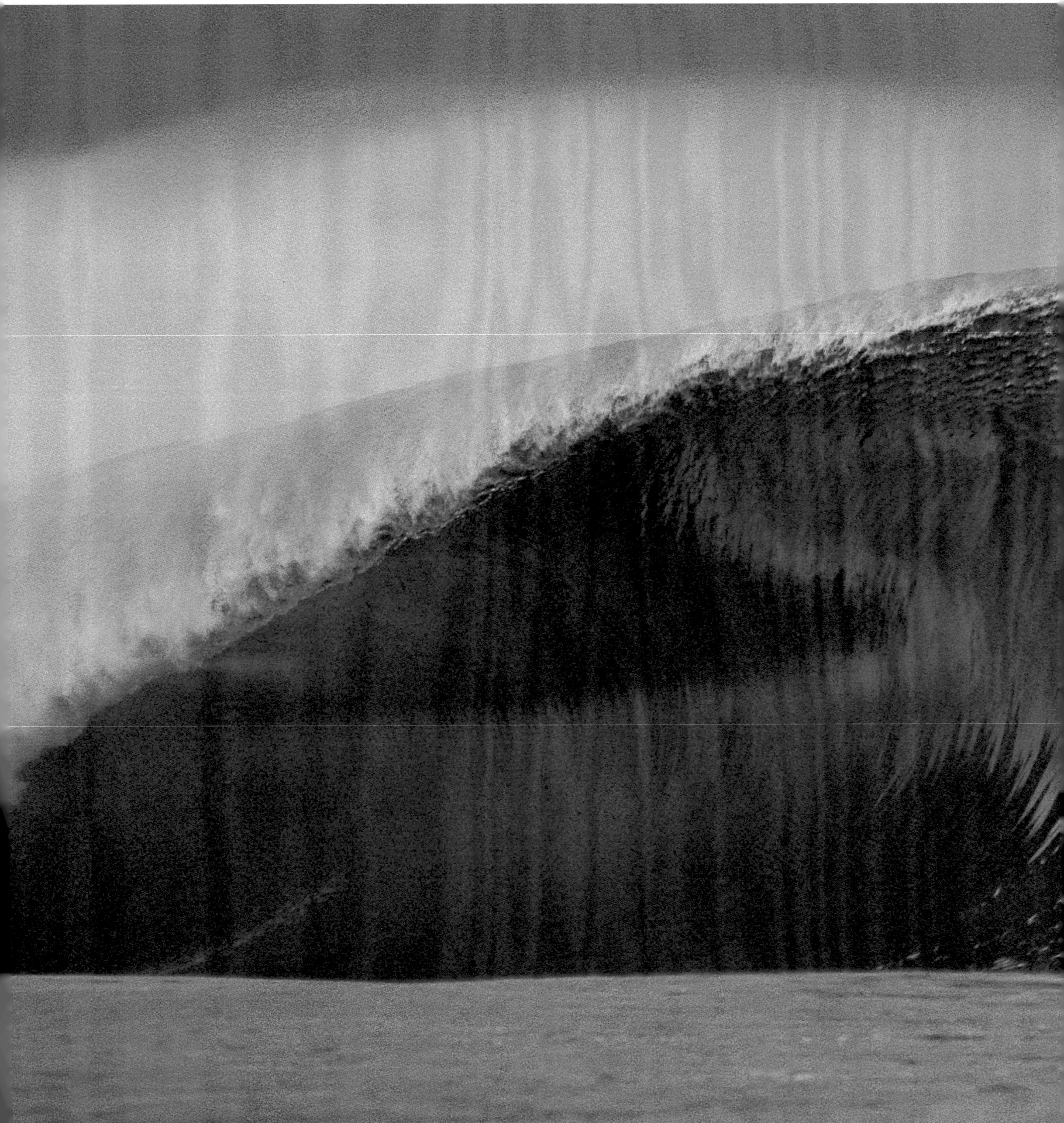

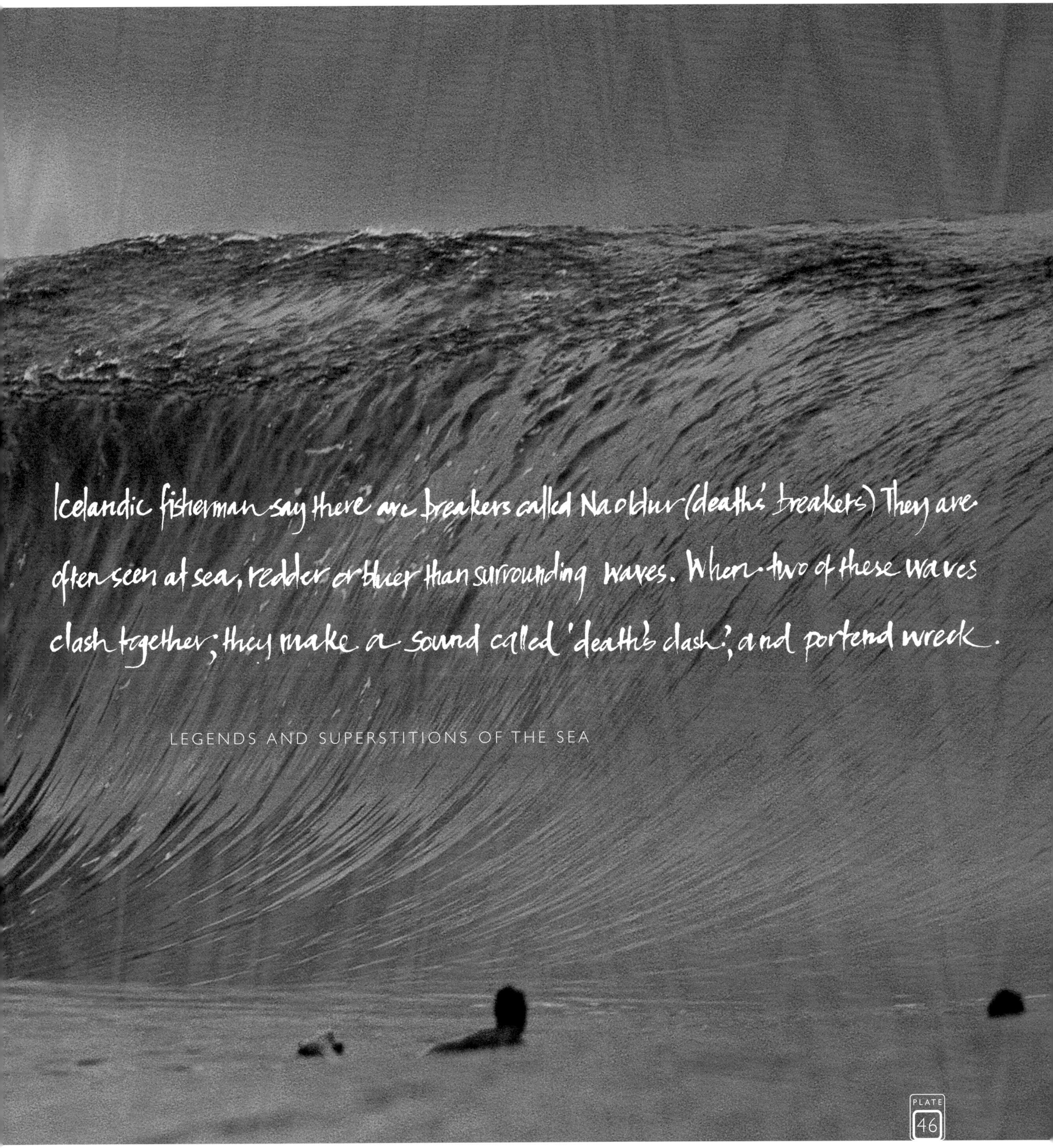
Icelandic fisherman say there are breakers called Naoldur (death's breakers) They are often seen at sea, redder or bluer than surrounding waves. When two of these waves clash together; they make a sound called 'death's clash', and portend wreck.
LEGENDS AND SUPERSTITIONS OF THE SEA
PLATE
46

PLATE **41** A frozen moment expresses the constant confrontation — ocean versus land. Numerous writers, artists and other observers of nature at work have seen parallels between this coastal battle and that of human nature; yet the sea wages its endless feud according to observable physical laws, and cares nothing for time in the process.

PLATE **42** The dynamic curve of a cresting wave is clearly seen from above as it approaches breaking point. It's possible to imagine the breaking wave as traffic on a freeway; suddenly the lower lanes slow, to be overtaken by the upper lanes, which pour past unimpeded. Of course, in the wave's world, gravity then takes a hand.

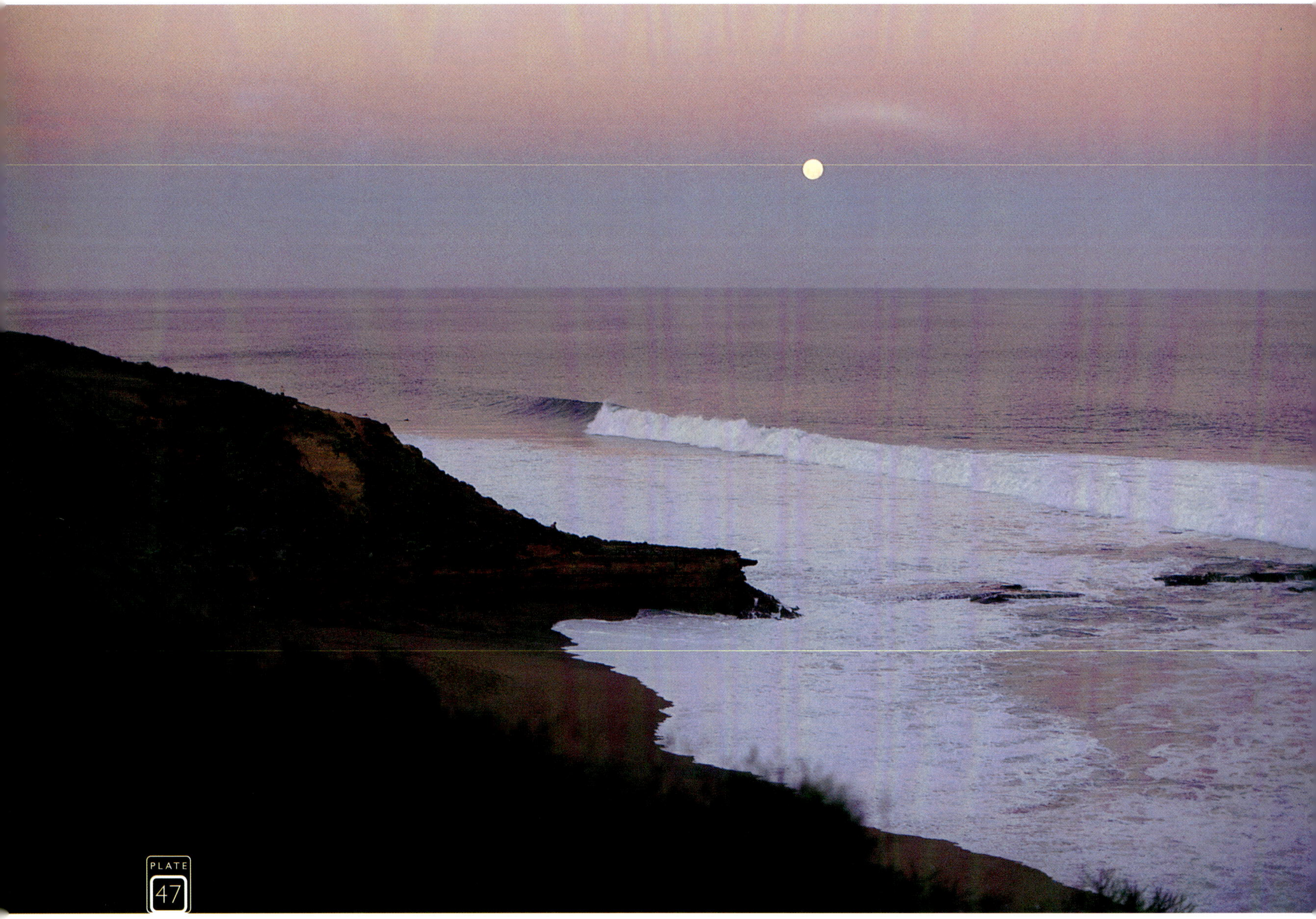

PLATE 47

PLATE **44** An insider's vision of the fabled "tube". To the surfer, trying to describe the sensation of riding within such a wave, surrounded by the turbine energy of the curl, calls into question the effectiveness of the language — any language. South African surfer Shaun Tomson has said, "Time is expanded in the tube", which is probably as close as anyone will get.

PLATE **45** In sudden resistance to the shore, a wave hurls its lip forth off Baja, Mexico. The lower half of the swell is brought to a stop so suddenly by a dramatically sloping shoreline, the plunging lip almost seems to detach itself from the swell in a mad effort to greet the sand.

PLATE **43** Swell lines are drawn this way and that under a stormy sky. Despite their ominous appearance, the storm clouds may have little to do with the swell that moves beneath; such long stretched lines almost certainly have their genesis in a storm fetch far from view.

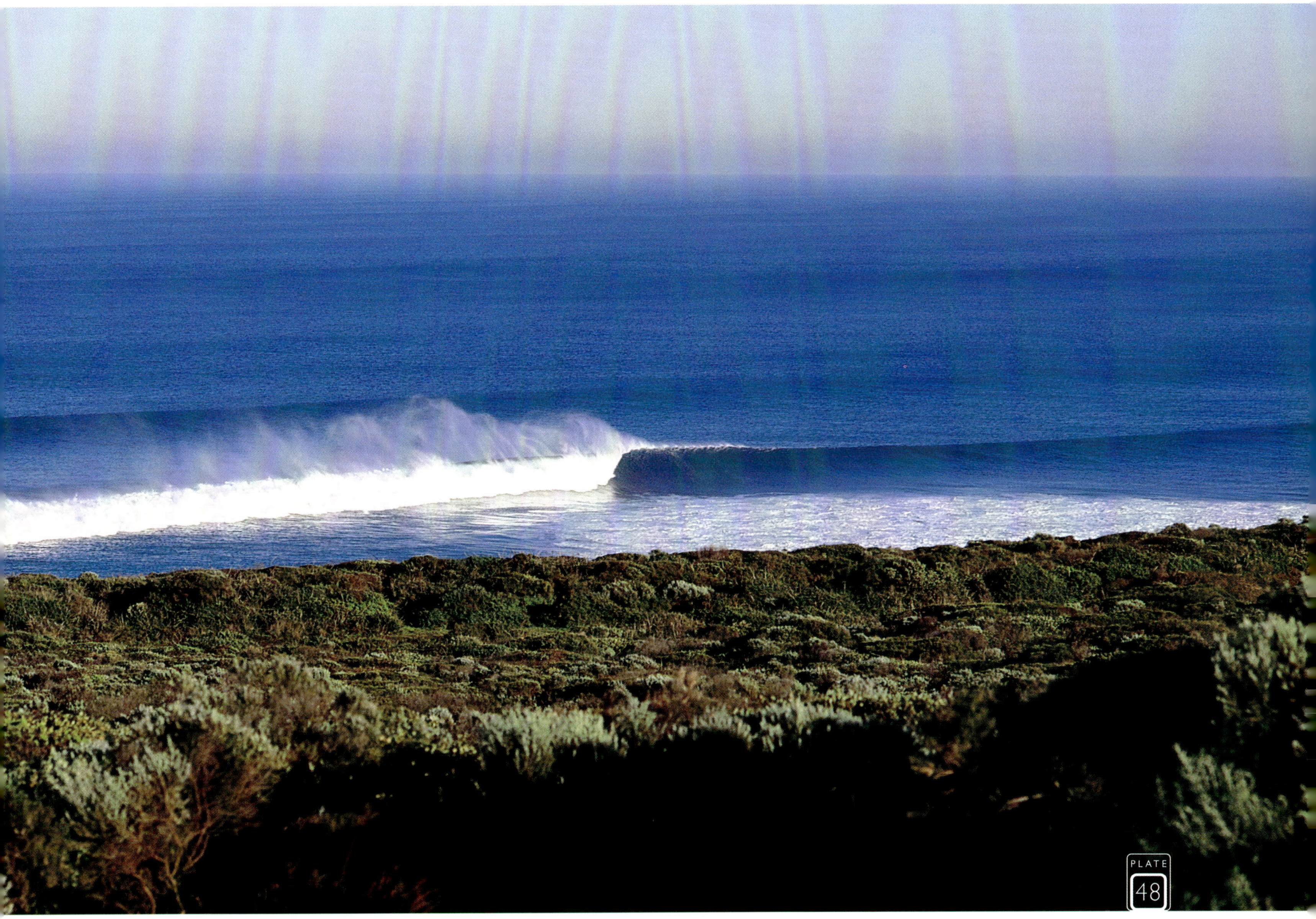

PLATE 48

PLATE **46** Like a black obsidian rock face, a wave stands up in ultimate challenge to the waverider at Teahupo'o, Tahiti. Such guillotining power confronts the surfer's every skill, forcing a split-second decision as to what can be ridden — and what can't.

PLATES **47/48** Two faces of a similar swell pattern. At left, Roaring Forties surf rolls gracefully down the reef at Winkipop, Victoria; at right, the same thing hits with a little more vigour two thousand kilometres away off the West Australia coast. It's possible for the keen surfer to ride waves at the spot on the right, climb in a car, and two days later catch surf from the same storm at Winkipop, as the swell moves endlessly around the southern fringes of the Australian continent.

There's a young man riding
on the wave-crest,
Now he's rising to the sky—
The sun! The foam!

Ruahie!
The wave folds down,
It's flat,
It's lost its force.

Easter Island

CHANT TO DUMP A SURFER

before 1934

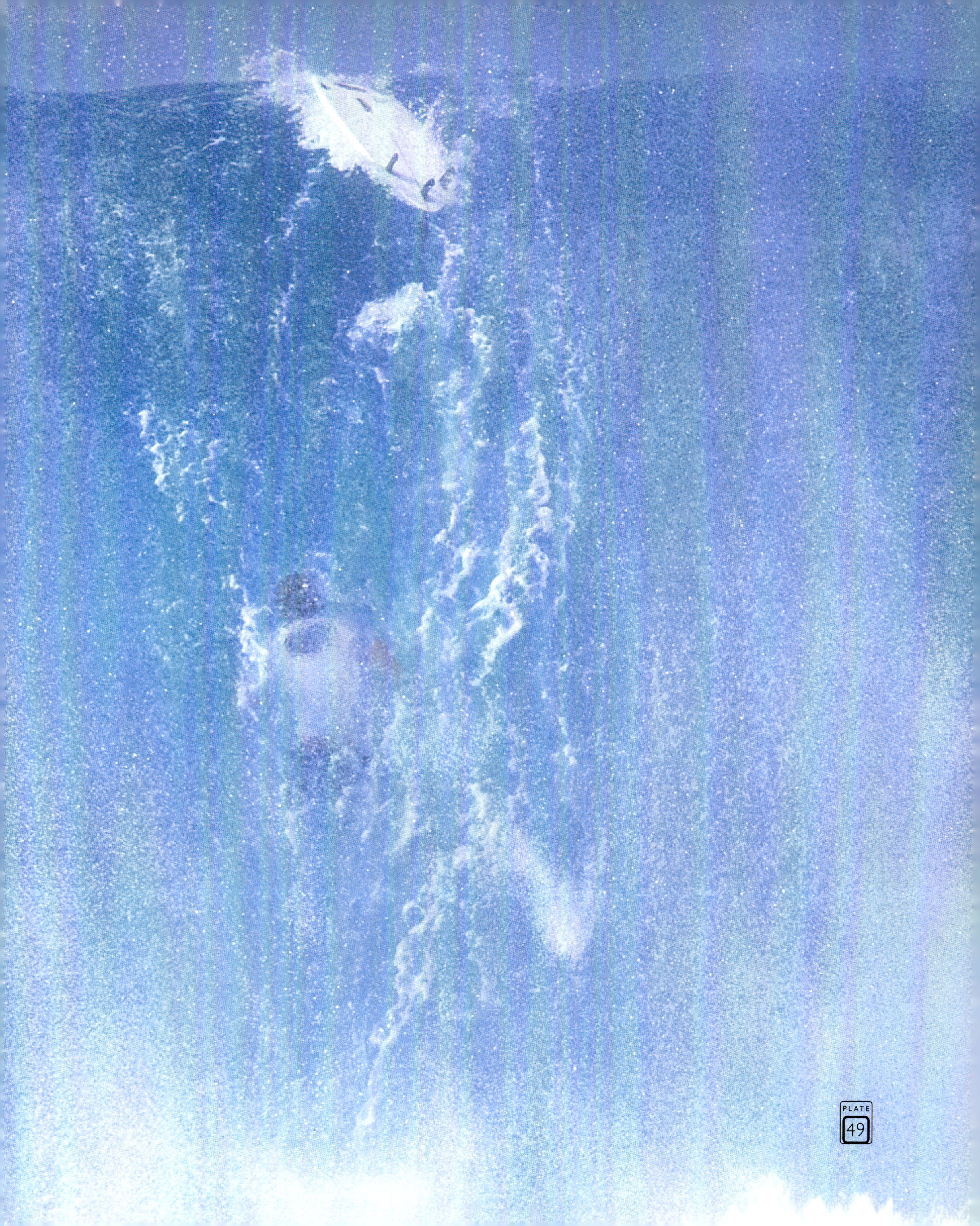

PLATE 49

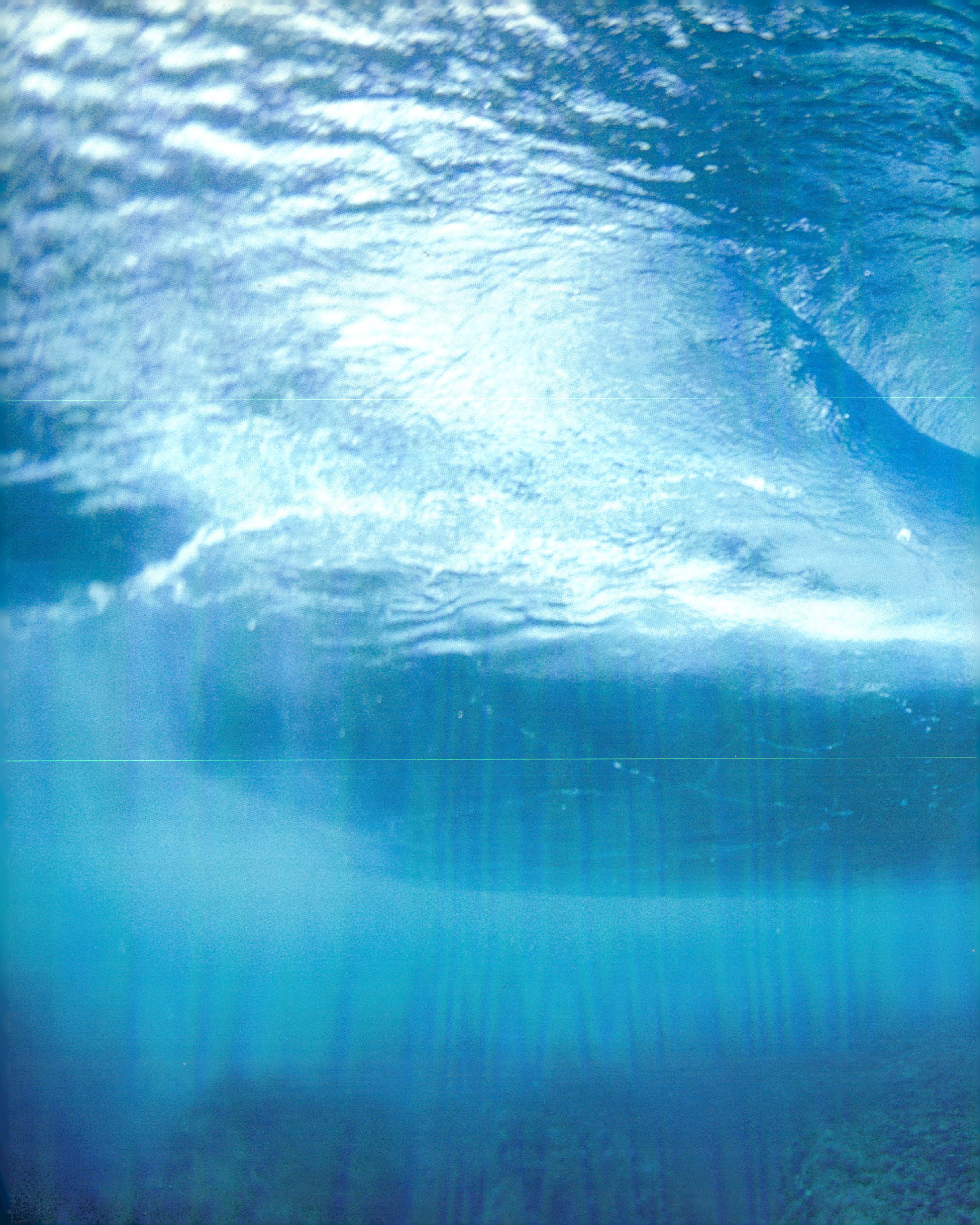

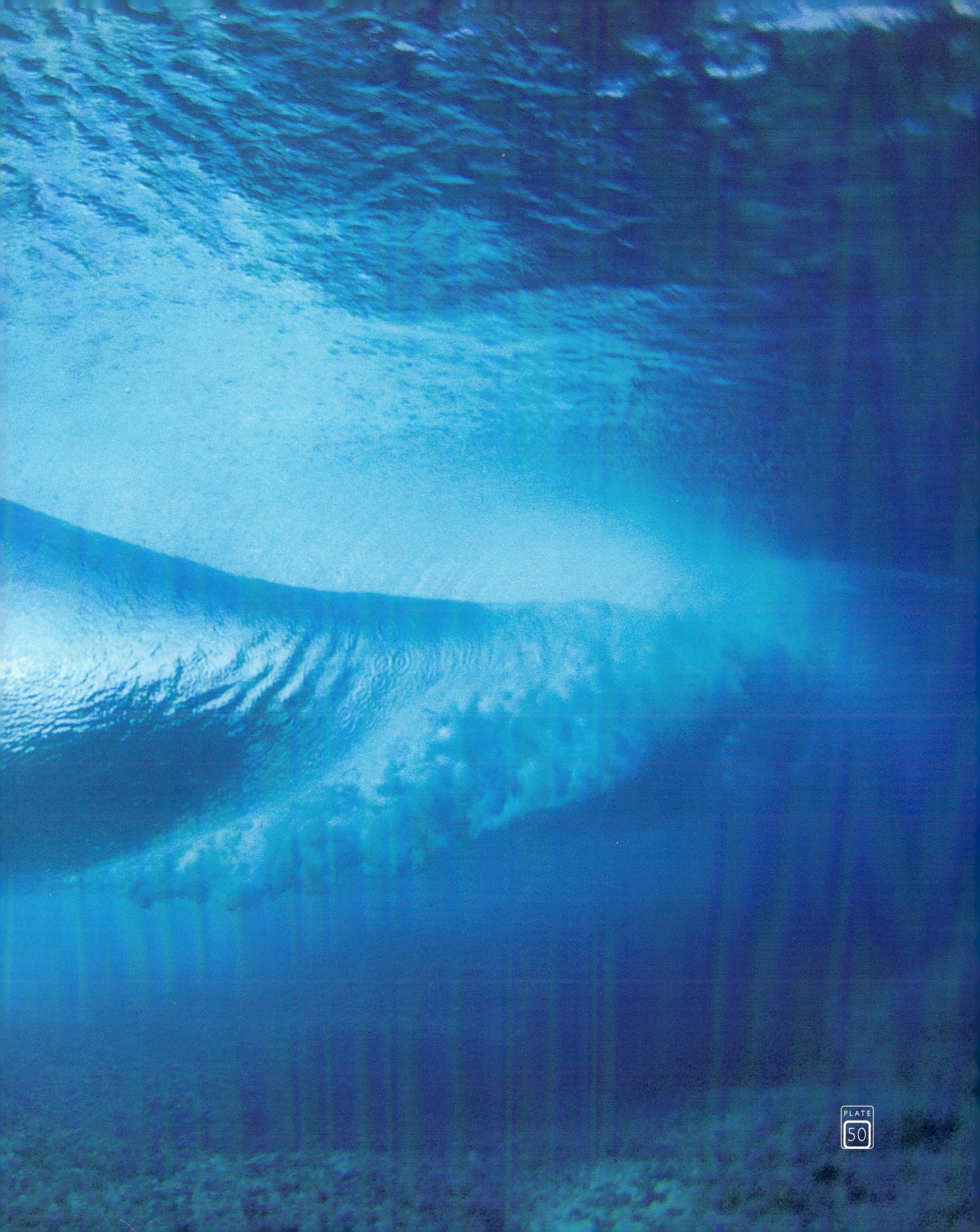

PLATE 50

As the set approached you could see the first wave jacking up on the inside bowl and the wind trying to hold it back. The beast stood up easily with a sixty foot face. With chops pushing up the face you could barely notice the warble rising within the wave. What I saw next scared me more than anything I've ever seen. When the warble got a little past halfway up, it started to crest and the monster doubled up. The thick lip just threw out with brutal force and the bottom dropped out at the same time. For those few seconds the wind, the water, and everyone's screams went silent. During those long seconds I looked down over my shoulders and could see the lip flying so far out and the bottom inverting about ten feet down. The silence was broken with a thundering crash as the lip exploded on the water. Even though I was safe, I flinched. At that moment, I realised how heavy it really was and that there are waves that are not meant for man... It was so ugly and at the same time so beautiful.

Ken "SKINDOG" COLLINS SURFING MAGAZINE april 1998

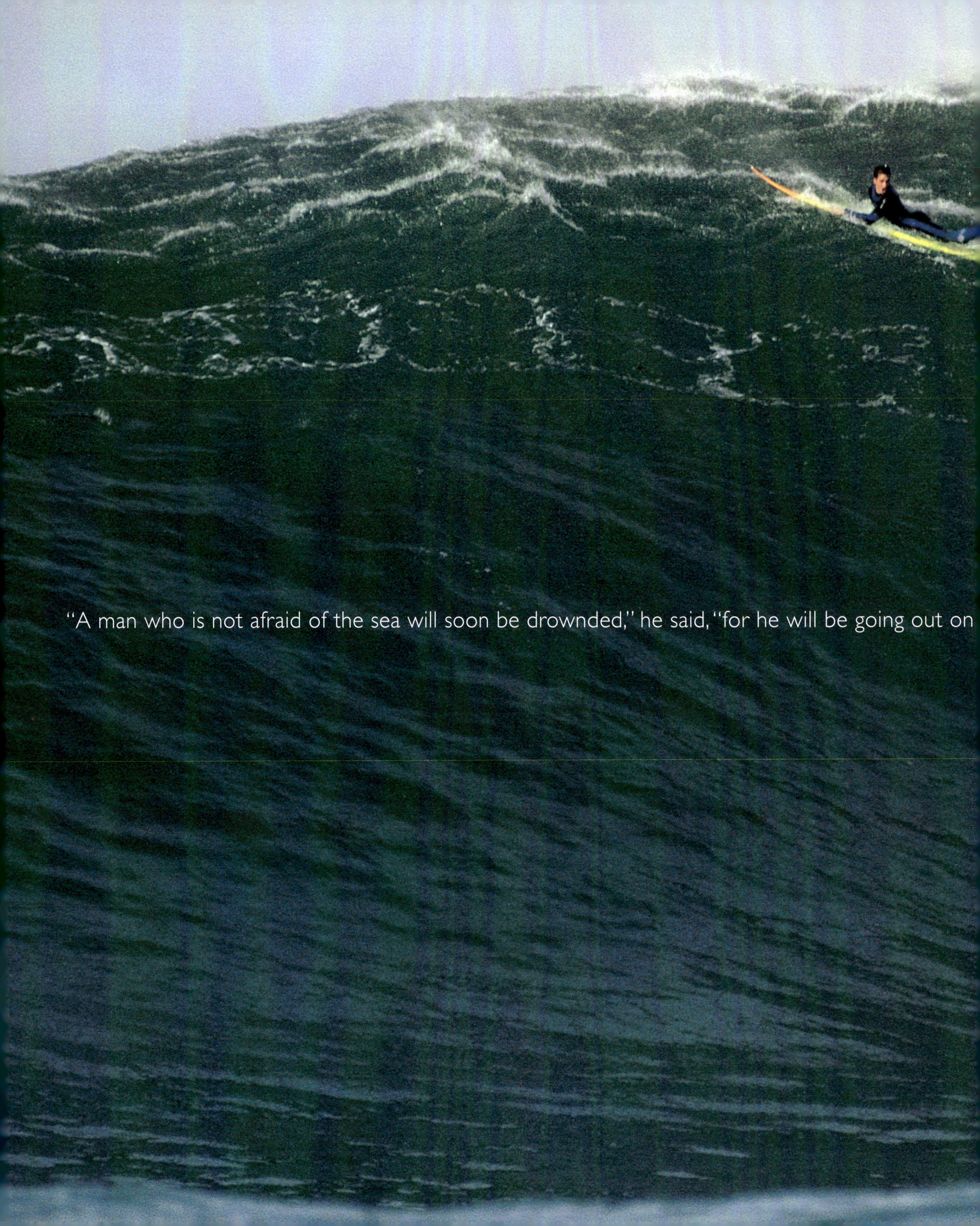

"A man who is not afraid of the sea will soon be drownded," he said, "for he will be going out on

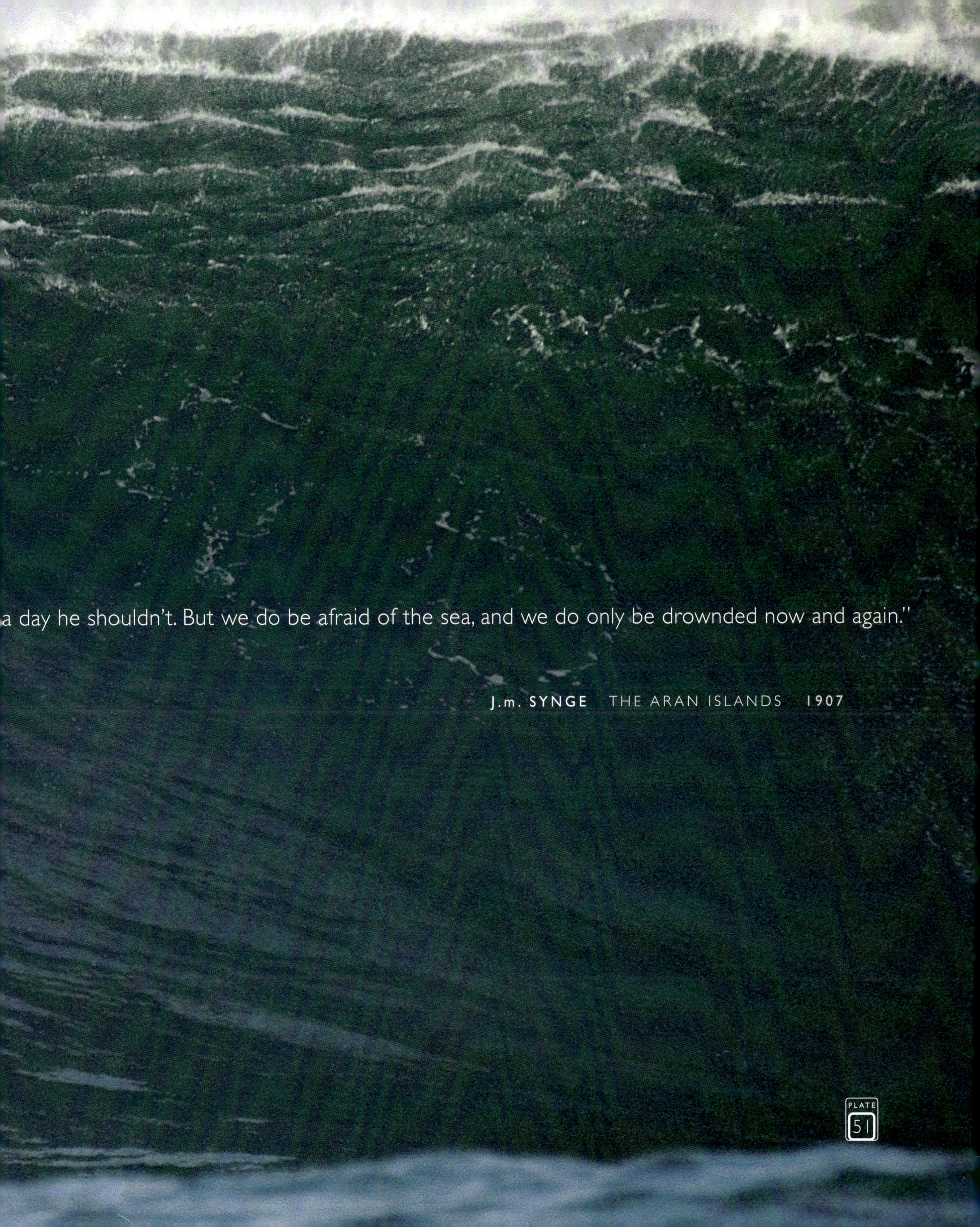

a day he shouldn't. But we do be afraid of the sea, and we do only be drowned now and again."

J.m. **SYNGE** THE ARAN ISLANDS **1907**

PLATE 51

The Sea hates a Coward!

Eugene O'NEILL MOURNING BECOMES ELECTRA 1931

PLATE **49** "Being one with the sea" takes on an entirely new meaning for Hawaiian surfer Andy Irons, as he's hurled down with a plunging lip at the Pipeline, Oahu. Once caught up in such circumstances, there's little for him to do but curl up and hope for the best. Pipeline regularly tops the worldwide surf location injury count each year; one day in early 1999 saw thirty separate bloodlettings, several of them serious enough to require helicopter evacuation.

PLATE **50** Seen from beneath, the invisible hand of swell energy as it works on water becomes ever more apparent, sculpting an intricate wave-form across a coral-lava reef. Such forms possess a curious consistency throughout the natural world, from this wave's curl to the swirling movements of tornados, cyclones, and even — on an unimaginably massive scale — the galaxies.

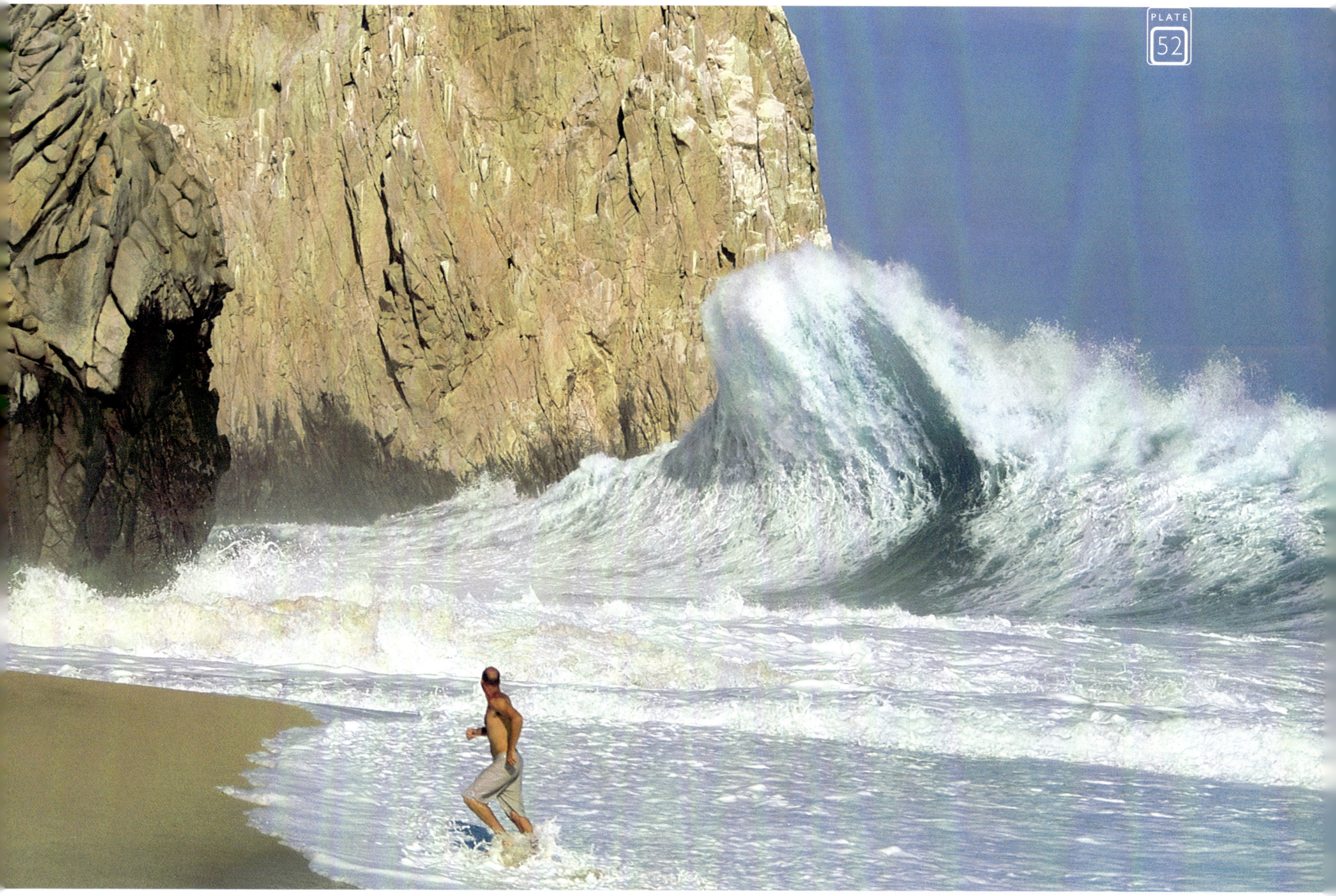

PLATE **51** Surfer Ken Collins takes a quick glance over his shoulder into the dark throat of a huge wave at Maverick's, the renowned surf spot near Half Moon Bay, California. The power of a wave increases as a product of the square of its height; the impact of this wave's lip as it lands will likely approach four tons per square foot.

PLATE **52** Reflection magnifies a breaking wave near Cabo San Lucas, Mexico. It's a well-known phenomenon among surfers: swells bounce off a cliff or jetty and effectively multiply their breaking height, sometimes resulting in great rideable surf — though not in this case.

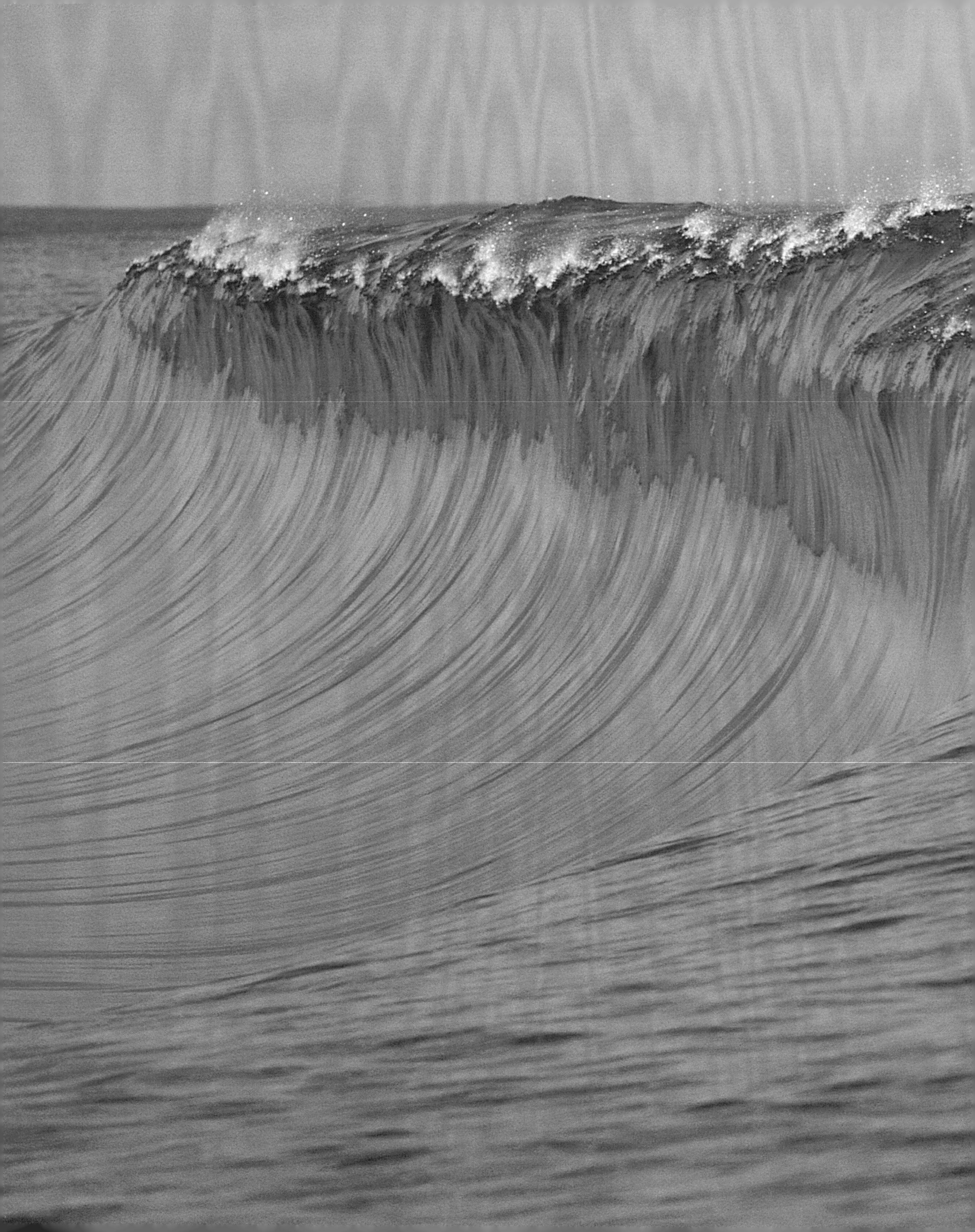

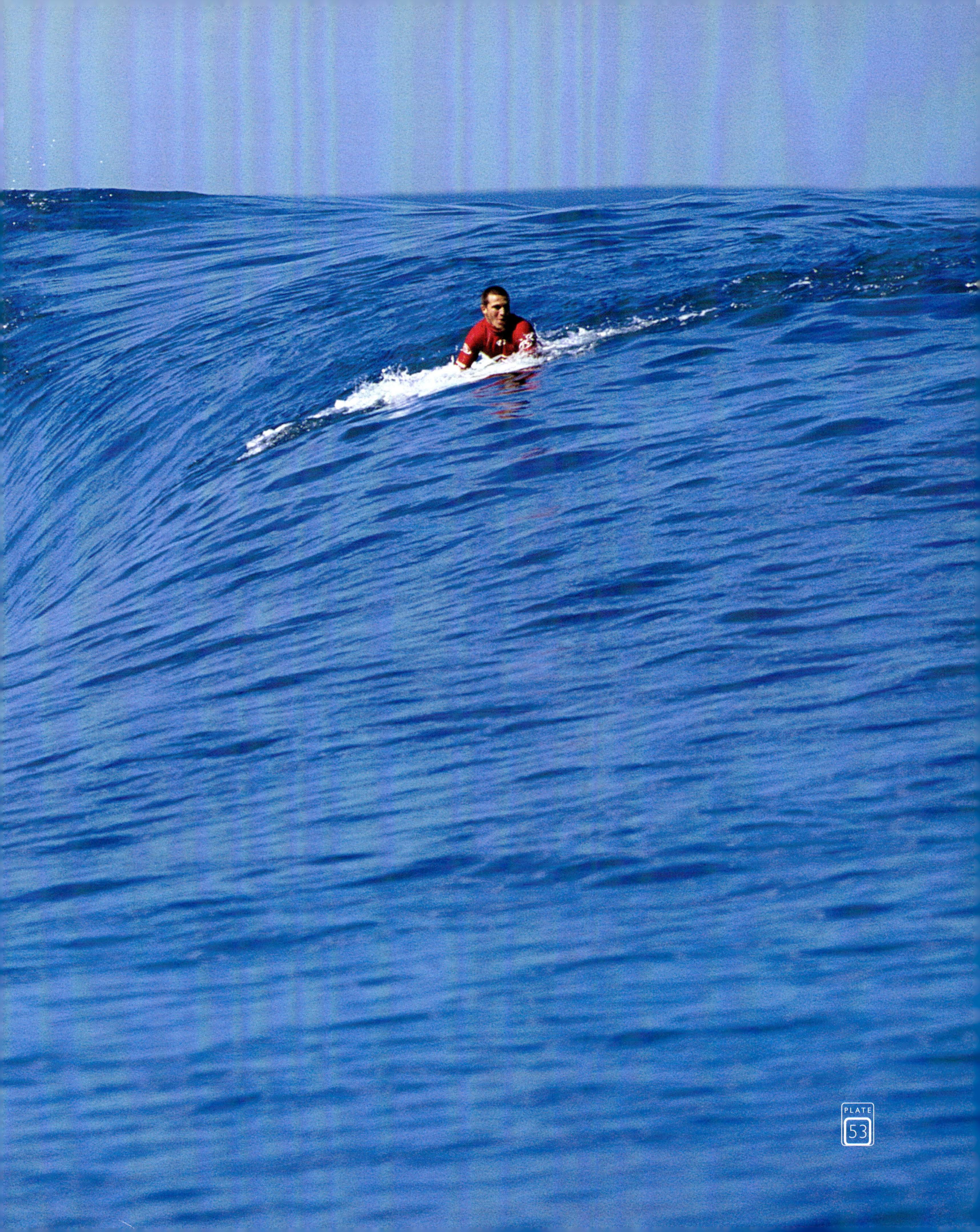

PLATE 53

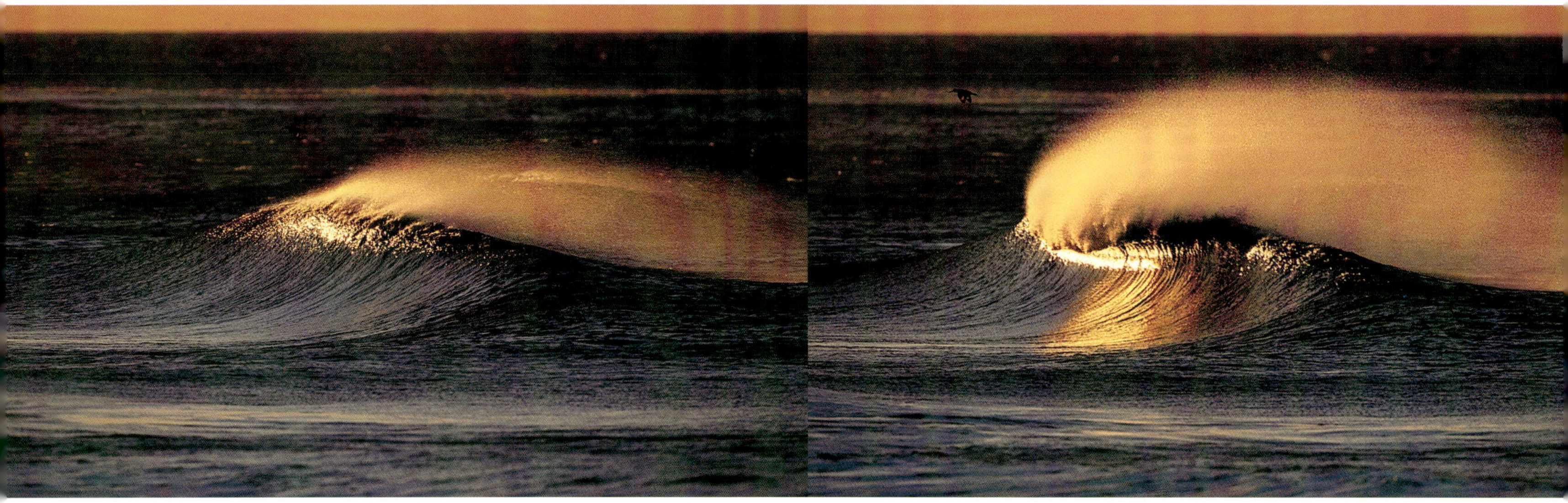

An observer sitting on a beach can make at least an intelligent guess whether the surf spilling

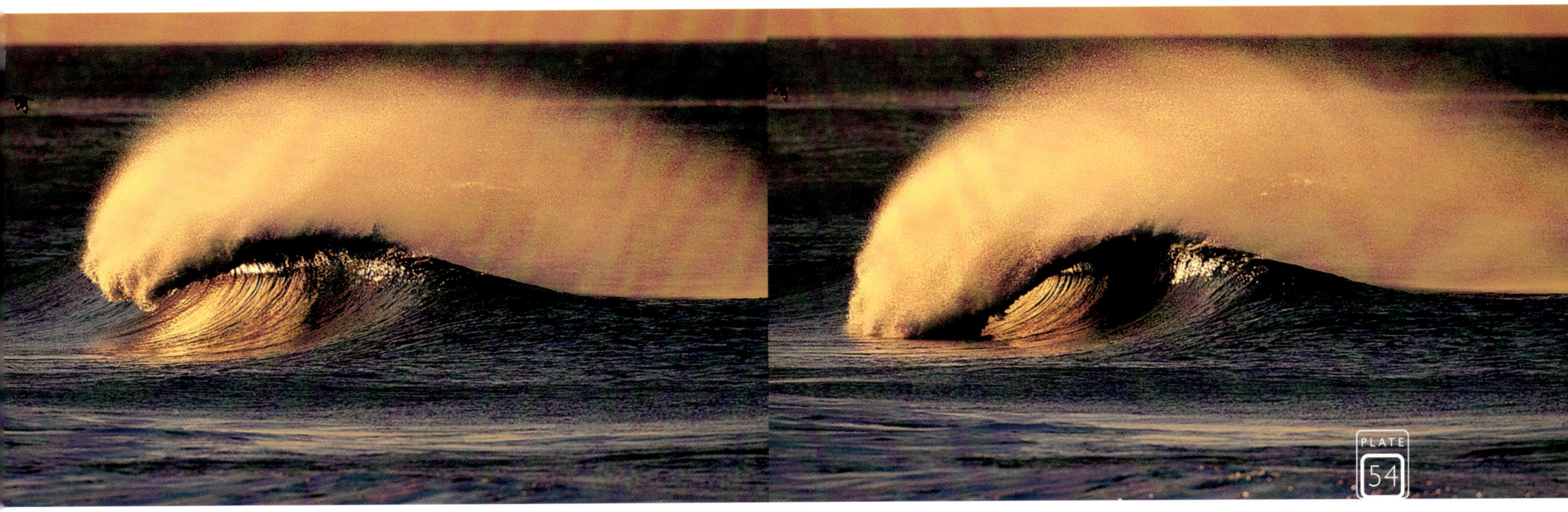

out onto the sand before him has been produced by a gale close offshore or a distant storm ...

... Young waves, only recently shaped by the wind, have a steep, peaked shape even well out at sea. From far out on the horizon you can see them forming whitecaps as they come in; bits of foam are spilling down their fronts and boiling and bubbling over the advancing face, and the final breaking of the wave is a prolonged and deliberate process. But if a wave, on coming into the surf zone, rears high as though gathering all its strength for the final act of its life, if the crest forms all along its advancing front and then begins to curl forward, if the whole mass of water plunges suddenly with a booming roar into its trough — then you may take it that these waves are visitors from some very distant part of the ocean, that they have travelled long and far before their final dissolution at your feet.

Rachel CARSON THE SEA AROUND US 1951

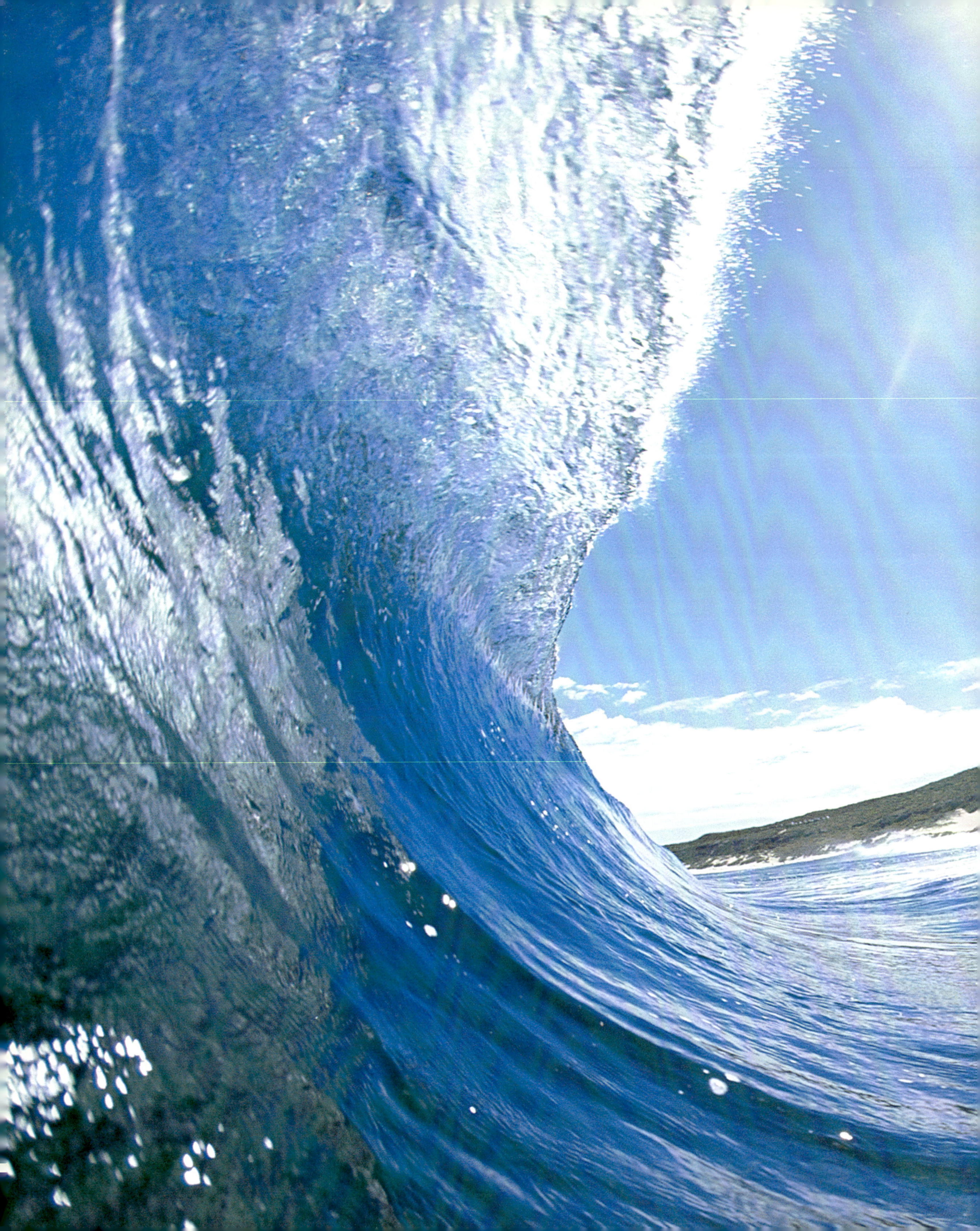

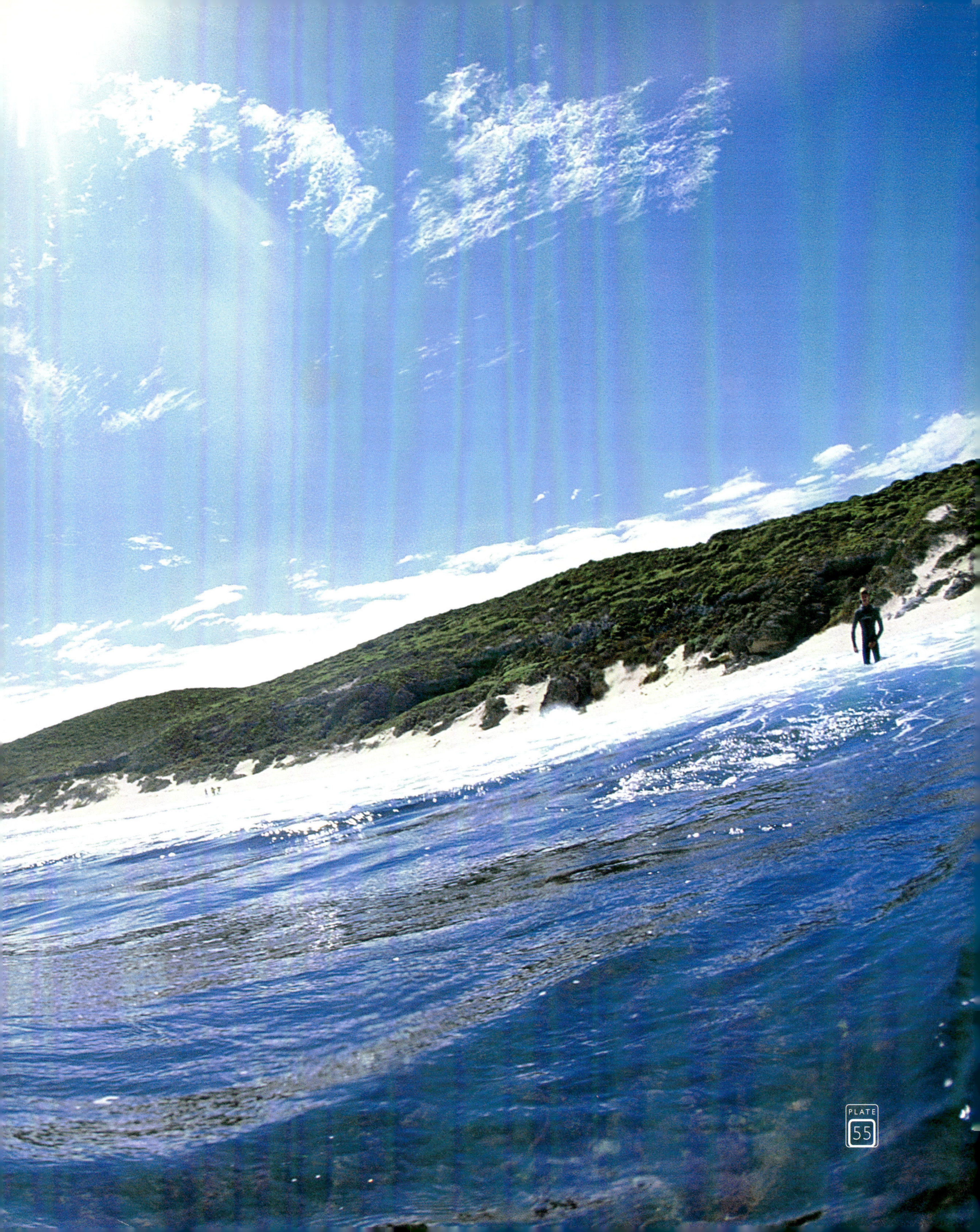

PLATE 55

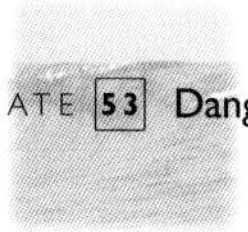

PLATE **53** Danger lurks in the lower left of this seemingly perfect wave in the South Pacific — a shallow coral reef, almost exposed by the swell's dramatic draw. An incautious surfer might risk a painful, possibly serious pounding; studying and marking the danger points is an important part of tackling such breaks for the first time.

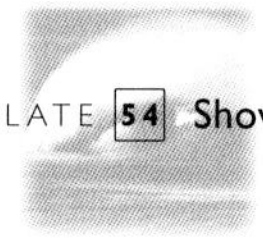

PLATE **54** Showing the flawless texture typical of a smooth sand bottom contour, a small wave demonstrates what may be the key to humanity's fascination with the shore — a beautiful sculpture that with each passing moment, alters its shape to something equally beautiful.

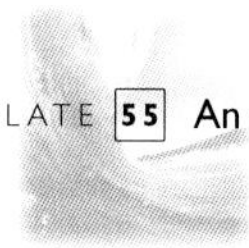

PLATE **55** An explosive swell shows its form on a remote West Australian beach, while a cautious surfer waits for his chance to paddle out. Timing is a surfer's greatest challenge — timing the water entry, timing the oncoming swell's intersection with the reef or sandbar, timing the ride to coincide with the wave's movement toward the shore.

PLATE **56** Like a volcanic geyser, water shoots many metres into the air through a blowhole at Three Tables on Oahu. Blowholes are common on rocky coasts exposed to constant wave action, when water pressure wears into submerged rock strata and finds a weakness in the surface slab.

The sea — this truth must be confessed — has no generosity. No display of manly qualities — courage, hardihood, endurance, faithfulness — has ever been known to touch its irresponsible consciousness of power.

Joseph **CONRAD** THE MIRROR OF THE SEA 1907

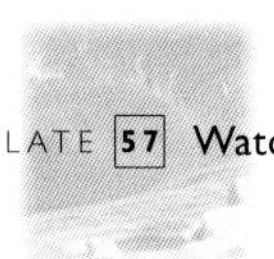

PLATE **57** Watched by a sensibly cautious surfer, outrageous energies are released by a potent South Pacific swell as it connects with shallow reef. Without doubt, some of such a vision's fascination for us is the fact that despite all the science and art at our disposal, understanding waves — really knowing why they exist in our physical universe — is still beyond us.

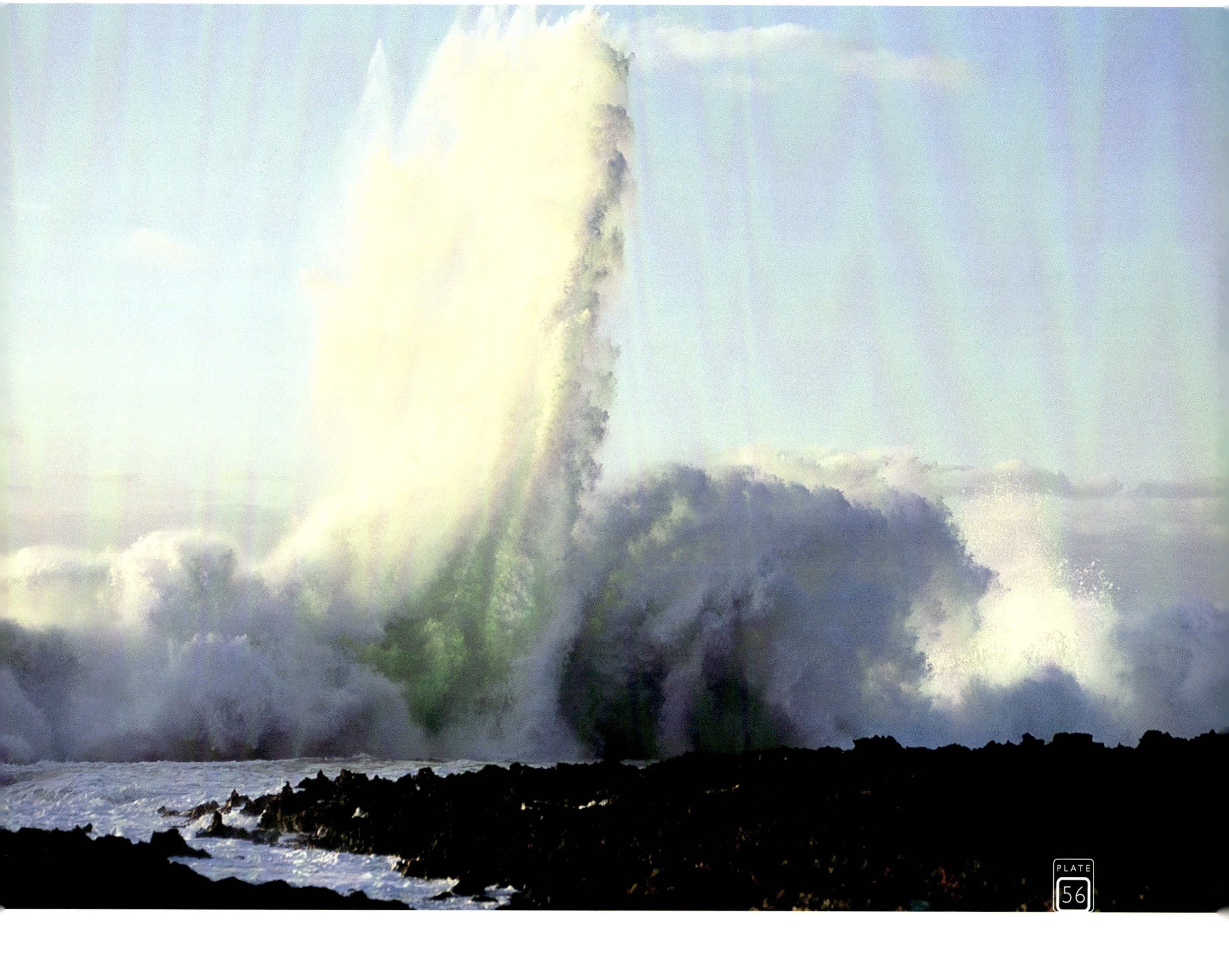
PLATE
56

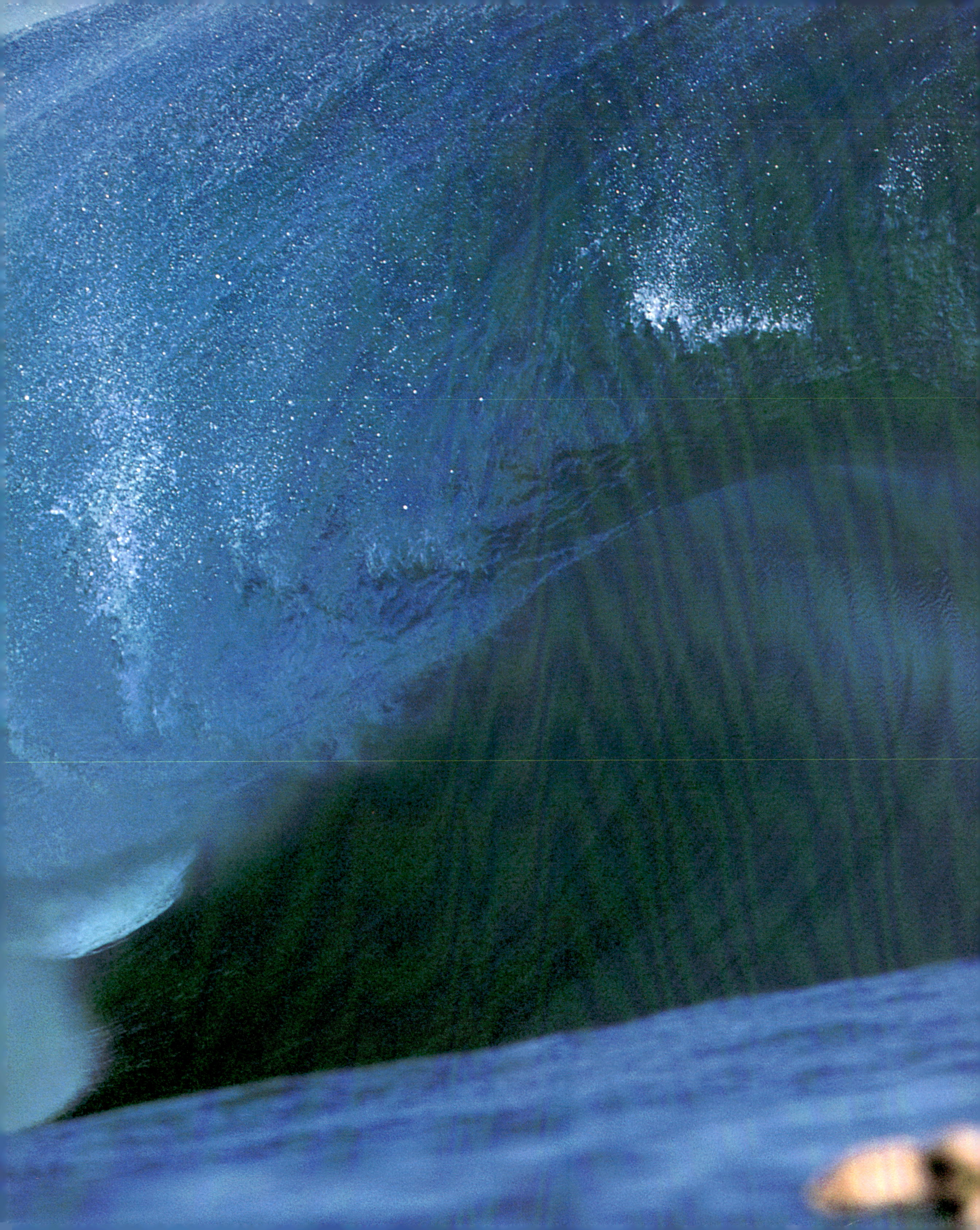

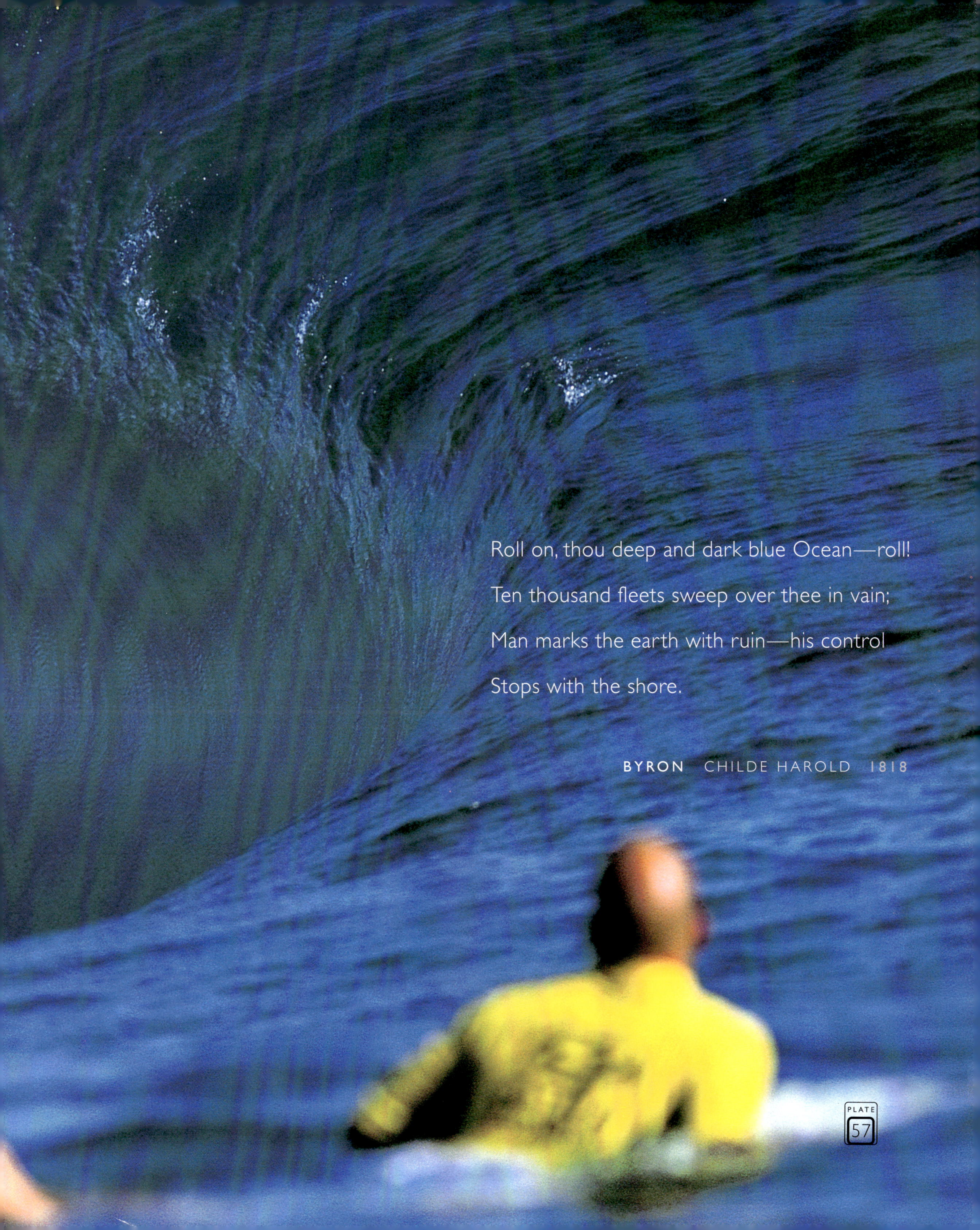

Roll on, thou deep and dark blue Ocean—roll!
Ten thousand fleets sweep over thee in vain;
Man marks the earth with ruin—his control
Stops with the shore.

BYRON CHILDE HAROLD 1818

PLATE 57

Here are some **REFERENCES AND CREDITS**

for excerpts contained in this volume

JIM BANKS, introduction from AUSTRALIA'S SURFING LIFE magazine, published 1995 by Morrison Media, Australia.

TIM WINTON, from AN OPEN SWIMMER, published by Allen and Unwin, 1982. Reprinted with permission

WILLARD BASCOM, from WAVES AND BEACHES: THE DYNAMICS OF THE OCEAN SURFACE, © 1980 by Willard Bascom.

LANGSTON HUGHES, from SEA CALM, first published by Alred A. Knopf, 1959.

RACHEL CARSON, from THE SEA AROUND US. Published by Oxford University Press, New York.

WILLIAM G. VAN DORN, OCEANOGRAPHY AND SEAMANSHIP. Published by Cornell Maritime Press, Maryland, USA.

KEN COLLINS, from ONE SIZE FITS ALL, SURFING MAGAZINE, published 1998 by Western Empire Publishing, USA.

RIPPLES ON A COSMIC SEA: THE SEARCH FOR GRAVITATIONAL WAVES, by DAVID BLAIR AND GEOFF MCNAMARA, Allen and Unwin, Australia.

LEGENDS AND SUPERSTITIONS OF THE SEA AND OF SAILORS, by FLETCHER S. BASSETT, Singing Tree Press, Detroit, USA.

HENRY BESTON, from THE OUTERMOST HOUSE, published by Henry Holt and Co. USA.

Photographers

This volume was published by MORRISON MEDIA

bookshop distribution (australia)
DENNIS JONES & ASSOCIATES PTY LTD
PH: (03) 9720 6761 FAX: (03) 9720 4472

newsagent distribution (australia)
NEWSAGENTS DIRECT DISTRIBUTION
PH: (02) 9353 9911

bookshop distribution (usa)
WORLD OCEAN TRADING CO
PH: (805) 684 9591 FAX: (805) 684 5575

new zealand distribution
GORDON & GOTCH
PH: (09) 625 3000

FIRST PUBLISHED 1999

to order extra copies, contact MORRISON MEDIA
FREECALL 1 800 069 918 FAX: (07) 5576 1527
EMAIL moradmin@morrisonmedia.com.au

this volume was printed by TOPPAN PRINTING CO., (H.K.) LTD.
scanning and colour separations were done by NUSCAN (australia)

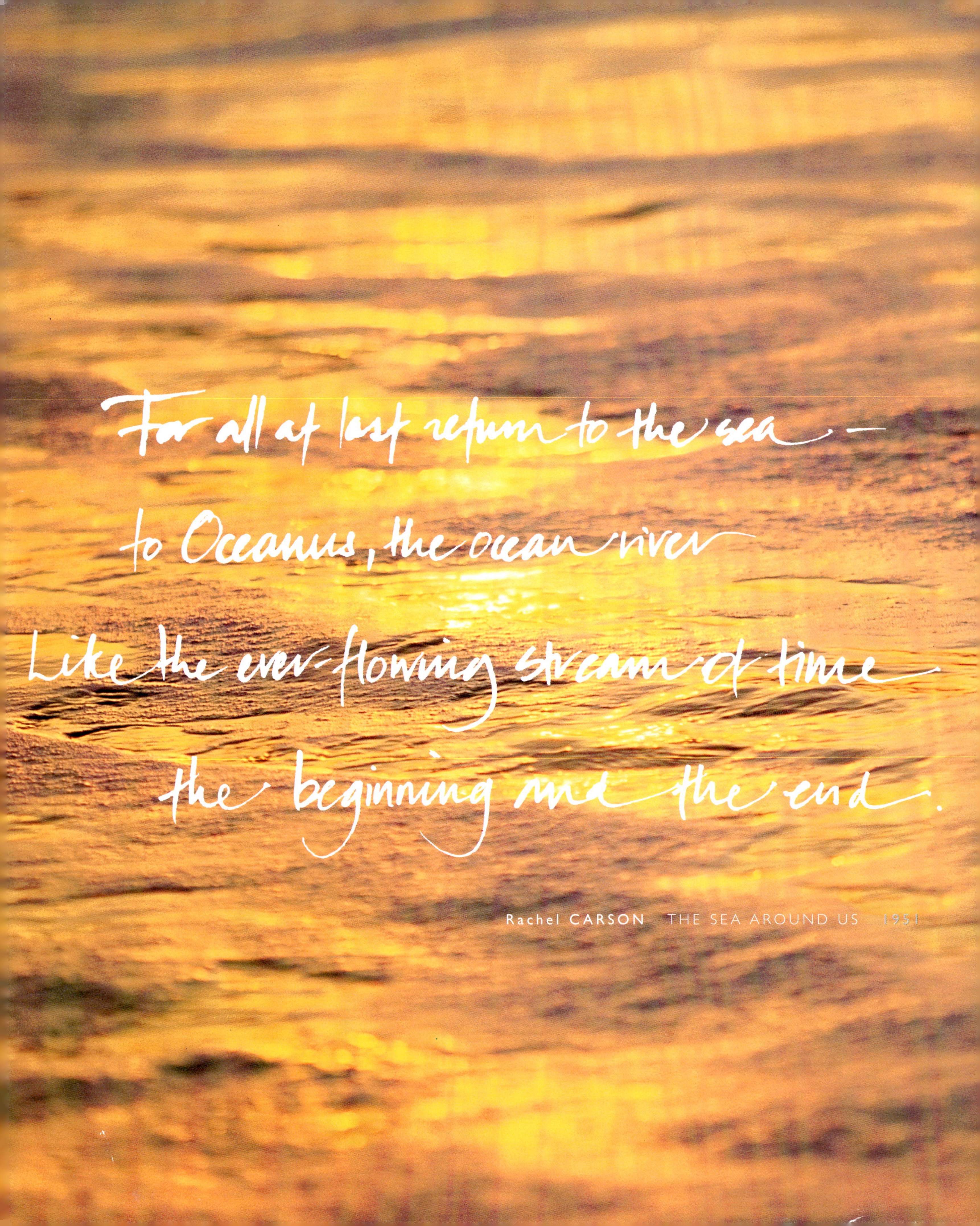
For all at last return to the sea —
to Oceanus, the ocean river
Like the ever-flowing stream of time
the beginning and the end.
Rachel CARSON THE SEA AROUND US 1951